US Security Issues and World War I

Edited by
Craig Greathouse and Austin Riede

UNG
UNIVERSITY of
NORTH GEORGIA
UNIVERSITY PRESS

Copyright © 2023 by Craig Greathouse and Austin Riede

All rights reserved. No part of this book may be reproduced in whole or in part without written permission from the publisher, except by reviewers who may quote brief excerpts in connections with a review in newspaper, magazine, or electronic publications; nor may any part of this book be reproduced, stored in a retrieval system, or transmitted in any form or by any means electronic, mechanical, photocopying, recording, or other, without the written permission from the publisher.

Published by:
University of North Georgia Press
Dahlonega, Georgia

Printing Support by:
Lightning Source Inc.
La Vergne, Tennessee

Cover design by Corey Parson.
Book design by Corey Parson and Ashley Topham.

ISBN: 978-1-959203-04-9

For more information, please visit: http://ung.edu/university-press
Or e-mail: ungpress@ung.edu

Table of Contents

Introduction   v

"We are Getting into Deep Water Needlessly": From Neutrality to (Almost) a Third Anglo-American War in 1914   1
Shawn McAvoy

Xenophobia Unleashed: Anti-German Attitudes and Policies During WWI   25
B.D. Mowell

American and Russian Geopolitical and Geostrategic Interests During World War I   52
Raluca Viman-Miller

"American Pride Will Not Stand in the Way of Efficiency": Public Opinion on the Amalgamation of US Troops into Allied Armies in 1918   76
Terri Blom Crocker

Lying, Spying, and Right Defying: The Espionage Act of 1917 and the US Wartime Qualification to the Freedom of Speech   99
Ashlee Beazley

League of Nations Debate: Strategic Preferences of President Woodrow Wilson and Senator Henry Cabot Lodge   128
Seyed Hamidreza Serri

One Hundred Years On: The Shadow of League of Nations Failure on American Support for International Law   158
Jonathan S. Miner

Cruelty or Military Necessity: Poison Gas and US Security in WWI   187
Thomas I. Faith

The AEF in the Trenches: The American Military and Modern Warfare in the First World War   211
Jonathan A. Beall

Proximity and Distance on the Battlefield: The AEF's 2nd Infantry Division at Blanc Mont 1918   238
Keith D. Dickson

AEF Press Censorship During World War I   263
Charles Sorrie

Contributor Biographies   289

Introduction

This project was initially proposed to address the 100th anniversary of America's involvement in World War I. However, it was not designed just to look at the military events that had occurred or foreign policy leading up to the war. Rather, the project was to consider the broader impact of the war on security issues within the United States and on US foreign policy. Additionally, the book was meant to incorporate a broad range of disciplines in looking at the history of the United States and World War I. Many works on World War I tend to focus on the fields of military history or international relations, which at times may limit our understanding of the influences and ideas that affected the United States leading into the war and coming out of the war. Therefore, the book was intentionally developed to allow other disciplines to engage with ideas and themes both domestic and international for American involvement in World War I.

For more than a century, the historiography of the US role in World War I has largely been about military strategy, and, to paraphrase Thomas Carlyle, the history of the war has also been the history of "great men." Naturally, the vast majority of these historiographies, including those recently published in commemoration of the war's centennial, have focused on what was happening in Europe. The changes to the security policies of belligerent European nations have necessarily been a part of these historiographies. Great Britain's Defence of the Realm Act, for example, is a well-known piece of authoritarian legislation passed immediately upon the outbreak of war, clearly a reaction to what appeared to be an imminent physical threat to British security. Meanwhile, in the United States, the apparatuses of national security had years to prepare for what was largely a hypothetical threat, as involvement in the war was not a popular notion. One of Woodrow Wilson's slogans for reelection was, "He kept us out of war." But what did the gradual mobilization of American security measures affect? And how did the military and civilians react, not just as the United States entered the war but during the nation's involvement for a year and a half? And what permanent or temporary changes to US security policy resulted from World War I? These are a few of the questions that the essays in this volume address. The US involvement in World War I was an extraordinary reversal of an isolationist ideal, which had already begun to crack with the Spanish-American War and other unprecedented forays into international affairs. It was also the beginning point for a century of US global ascendancy. The security decisions made during World War I were early steps in the US negotiation of how its increasingly international presence brought increasing risk, both to the military engaged abroad and to US civilians.

The book is divided into two sections, the first of which focuses on domestic and foreign policy surrounding American involvement in the Great War. The second focuses on unique military aspects of the war. The initial section is arranged roughly chronologically with the chapters moving from earlier decisions leading into the war to discussions about American decisions coming out of the war, showing a broad range of policies that World War I forced the American government to address.

The first chapter by Shawn McAvoy looks at the background of neutrality and the American position set out by Woodrow Wilson. The idea of neutrality shaped the American experience going into World War I, and this chapter explores some of the aspects that frame other policy choices addressed in later chapters.

The second chapter by B.D. Mowell looks at the backlash against Germans within the United States leading into the period of World War I. While this topic has been addressed in other sources, most readers will be surprised about the reaction of Americans to those of German descent. German Americans are now seen as a core group within the American melting pot, not the usual target of ethnic discrimination, but the reaction to this group during World War I shows the hostility that American culture can foster toward internal ethnic groups during periods of conflict with their affiliated nations.

In the third chapter, Raluca Viman-Miller looks at the broader geopolitical understanding between Russia and the United States during this era. This chapter illustrates the complexity of the relations between the US and Russian experience as it moved through the Bolshevik Revolution, complicating the dynamics of how the United States could address Russia while still dealing with the broader issues of World War I.

The fourth chapter of the book by Terri Blom Crocker moves beyond the positions of Woodrow Wilson and John J. Pershing to examine the reaction of the American public toward the question of incorporating the American military into the Allied forces or staying independent. Given the unprecedented nature of World War I for the American public, their reaction to how and whether the American military should be integrated within or separate from the Allied military command demonstrates conflicts between isolationism and a desire for international cooperation. This chapter explores ideas and themes that were new within American society and had not previously been part of American political debate.

The fifth chapter, by Ashlee Beazley, looks at issues arising from the Espionage Act of 1917. The concern over how to protect military information and ideas is a continuation of the freedom of speech debate that came out of arguments stretching back to the Civil War. This chapter looks at continued debates about the legal positions surrounding what is viable and what is not.

The final two chapters in Section 1 both look at the roles of the League of Nations and the United States. These two chapters are diverse considerations of how the United States addressed the League of Nations. Seyed Hamidreza Serri takes a more data-driven analysis while Jonathan S. Miner focuses on a legalist approach. These competing approaches provide the reader with different perspectives on the American reaction to and interaction with the League of Nations and international policy issues coming out of World War I.

Section 2 looks at issues surrounding the American military deployment to France. The range of chapters includes examinations of poison gas, the impact of trench warfare, the 2nd Division at Blanc Mont, and the influence of censorship on the troops at the front line. As with the rest of the volume, these chapters look at a range of issues that have not been a strong focus in other works. The diversity of ideas provides a broader understanding of the issues the American Expeditionary Force (AEF) had to address during its time in France and that the American politic apparatus had to grapple with in the aftermath of war.

Both Chapter 6 by Seyed Hamidreza Serri and Chapter 7 by Jonathan S. Miner deal with the contentious issue of the League of Nations. Serri examines the United States' specific strategic plans and desired outcomes in negotiating the League of Nations as it was being formed. Miner takes a long view of how the US failure to join the League of Nations had a lasting impact on the American public and political views toward international law throughout the 20th century and beyond.

In Chapter 8, Thomas Faith highlights the use of poisoned weapons during World War I and the ideas and themes underlying them. Considering World War I's introduction of the large-scale use of chemical weapons, this chapter highlights a critical ethical question that at times is covered over: should an existing weapon be engaged just because it exists?

In Chapter 9, Jonathan A. Beall focuses on the AEF in the trenches of World War I. Given the nature of combat on the Western Front and the importance of trench warfare, this chapter examines the specific impact of trench fighting on the AEF. The experiences of the AEF from the French and British positions in terms of trenches is an important dividing line when dealing with the history of World War I, and this chapter helps to highlight the distinctions between AEF and British trench experiences.

In Chapter 10, Keith D. Dickson looks at the ideas of proximity and distance in relation to the 2nd Infantry Division and shows the difference in mindset between the AEF and how other states were approaching the war by 1918. Given that the US military came into the war with some initial lessons from Europe's previous 3 years of fighting, the differences in tactics employed by the AEF are important to show the divergence in lessons learned.

The last chapter of the book by Charles Sorrie moves away from traditional military tactics and examines the issue of press censorship in Europe during the AEF's deployment. Given the nature of how information moved during this time, letters from the troops and other communications by the AEF could provide important guidance for the enemy. Thus, the use of censorship to control information was of critical concern. This chapter engages in a deeper discussion about why the control of information mattered and how censorship worked to counter the passage of important information.

These 11 chapters represent a different and distinct look at American security issues leading up to and coming out of World War I. They address some topics that have been considered in detail elsewhere, but many provide unique ideas and themes that have not been the focus of other detailed studies. The diversity of these chapters helps to expand our understanding of the unique ideas and themes that affected the United States as it engaged in World War I. The editors wish to thank the authors of these unique takes for their work and patience through the process of bringing this book to print.

1

"We are Getting into Deep Water Needlessly": From Neutrality to (Almost) a Third Anglo-American War in 1914

Shawn McAvoy
Patrick & Henry Community College

Abstract

In 1914, President Woodrow Wilson's efforts vis-à-vis the 1909 Declaration of London reflected the interests of American industry, which drove the president to push the Declaration as neutral law upon the belligerents in the Great War. This in turn led Wilson to pronounce the British blockade of the Netherlands illegal. Wilson further complicated the situation by convincing himself that the crisis over the Declaration paralleled the crisis between the United States and Britain which led to the War of 1812. Not understanding international law as well as others in his administration, such as the US Ambassador to Great Britain Walter Hines Page, Wilson stubbornly ignored advice that the United States could make major concessions to Great Britain regarding the Declaration yet remain neutral according to international law. In the event, Britain refused to accept Wilson's interpretation of neutrality. As a result, the Anglophile Wilson almost found himself unintentionally allying with the Central Powers against Great Britain.

Keywords: Neutrality, Declaration of London, World War I, Woodrow Wilson

"We are Getting into Deep Water Needlessly": From Neutrality to (Almost) a Third Anglo-American War in 1914

As war engulfed Europe in summer 1914, the United States watched in horrified fascination. Woodrow Wilson, a president who had entered the White House in 1913 with only 41.9% of the popular vote in a four-way race, steered a neutral course for the country (Gould, 2008, pp. 176, 192). As one path to this neutrality, Wilson agreed to support the London Declaration of 1909 as an international code of naval warfare and neutral rights.

In 1909, ten signatory nations, including Germany, France, Japan, Russia, and the United States, had intended the London Declaration concerning the Laws of Naval War, to give it its full name, to become international law governing naval warfare (ASIL, 1915a, p. 199). The Declaration provided for three categories of goods in warfare: absolute contraband, used exclusively for warfare, such as gunpowder, guns, and armor plates; conditional contraband, which a neutral could ship only if the materials would not become war material such as fuel, gold and silver, and foodstuffs; and non-contraband or free items, which many nations did not consider war related in 1909. Copper and rubber became the commodities of contention for London. Relegated to the free list in 1909, these commodities assumed important military applications by 1914 (Gregory, 1970, p. 64). Additionally, the submarine already disconcerted London during the 1909 negotiations since enemy submarines could prevent the Royal Navy from conducting effective blockades (Martin, 2009, p. 737).

The other nine signers awaited Britain's ratification before submitting formal ratifications. The British House of Commons accepted the Declaration in December 1911, but the House of Lords rejected it 5 days later. Wanting more material declared absolute, and less material free, London objected to the exact division of categories and refused ratification. As a result, none of the ten signatories ever formally ratified the Declaration even though some informally accepted it (ASIL, 1915a, p. 201). The US Senate recommended ratification on April 24, 1912, but President William Howard Taft refused to endorse it (Savage, 1936, p. 163).

Wilson's efforts vis-à-vis the 1909 Declaration of London reflected the interests of American industry. Wilson then complicated the situation by convincing himself that the crisis over the Declaration paralleled the crisis between the United States and Britain which led to the War of 1812. He stubbornly ignored advice that the United States could make major concessions to Great Britain regarding the Declaration yet remain neutral according to international law. In the event, Britain refused to accept Wilson's interpretation of neutrality, and the Anglophile Wilson almost found himself unintentionally allying with the Central Powers against Great Britain.

The Declaration in the Great War

When the Great War broke out in August 1914, Secretary of State William Jennings Bryan, a staunch pacifist, prevailed upon President Wilson to convince the belligerents to accept the Declaration of London as a code of neutral rights (Gregory, 1970, p. 28). American businesses wanted a definite list of materials which they could ship to neutral ports under threat of blockade. The undeclared British blockade of Europe endangered American profits, as the British navy eliminated the German merchant marine from the Atlantic (Gregory, 1970, p. 30).

Per the Declaration, for a belligerent to impose a blockade, that belligerent first had to announce the blockade publicly. Then the blockade had to be *de facto* and absolute. London knew it could not conduct an effective blockade against Germany. German submarines rendered any close blockade of the continent dangerous to the point of impossibility, and Britain had no way to blockade Germany in the Baltic where Swedish goods could reach Germany almost without harassment of any kind (Marsden, 1977, pp. 489-490). British plans for naval war against Germany dated back to at least 1912, including plans to confiscate contraband bound for Germany (Marsden, 1977, p. 488).

One major issue for Britain became the Dutch port of Rotterdam, which, per the Rhine Treaty of 1869 with Germany, had become a *de facto* German port for military supplies (Page, 1914a). Rotterdam technically was a neutral port, but all of the foodstuffs and other cargoes ultimately went to Germany. In an August 3rd meeting involving Foreign Secretary Sir Edward Grey, First Lord of the Admiralty Winston Churchill, Chancellor of the Exchequer David Lloyd George, and Prime Minister Herbert Asquith, the ministers decided against an Admiralty request to blockade Rotterdam since that might undermine London's claim to enter the war to defend "small nationalities." Churchill worked successfully to get the decision reversed (Lambert, 2012, pp. 205–206).

London's blockade reflected its war aims, aims which its allies did not share. France and Russia wanted victory over a partitioned Germany, but Britain planned for an Entente victory, which resulted in a territorially whole, but militarily stripped, Germany (House, 1914, p. 138–139). To achieve this goal, London unilaterally added to the absolute contraband list. The list, which included 12 items in August 1914, swelled to 21 by October (Gregory, 1970, p. 65). London, per the Declaration, had the right to expand the absolute contraband list, but free items had to remain free, which presented problems for London (Gregory, 1970, p. 68).

Wilson saw trading with all belligerents as necessary, not just for the US economy but to prepare the country for a possible entry into the war. As a result, he rejected the suggestion of Jacob Schiff, a banker with Kuhn, Loeb & Co., that the United States maintain strict neutrality by neither lending money to the belligerents nor selling arms and ammunition (House, 1914, p. 232). Kuhn, Loeb & Co. benefitted from the German war effort as one of the few international banks to fund Berlin until London's declaration of war on August 4th (Schuler, 2011, p. 86). Standard Oil, copper interests in the West, cattlemen in the Midwest, and cotton growers in the South all pressured Wilson regarding the Declaration (Lambert, 2012, p. 268). Wilson responded by pushing the Ship Registry Bill through Congress, which would allow foreign-built merchant ships, primarily those from Germany, to register as American (New York World, 1914, p. 324). British Ambassador Cecil Spring-Rice cabled Grey that William Gibbs McAdoo, secretary of the treasury and presidential son-in-law, was behind the Ship Registry Bill and stood to profit from it since Kuhn, Loeb & Co. brokered the sale of German ships caught in American ports at the war's beginning. The British seemed to share a Republican suspicion that McAdoo had ties to Kuhn, Loeb & Company. French sources confirmed the intelligence, and Grey told Spring-Rice to inform Wilson, but Spring-Rice decided against it (Lambert, 2012, pp. 243, 248; Gould, 2016, p. 32).

At this point, events temporarily removed Wilson from the scene. On August 6, 1914, First Lady Ellen Axson Wilson died (Link, 1960, p. 193). Wilson's main advisor, Edward M. House, had kept in contact with the First Lady's doctor and knew that Wilson had been unaware of the severity of his wife's condition, so her death hit him particularly hard (House, 1914, p. 138-139). Wilson withdrew from his duties and left foreign affairs to House. House was pro-British in action, but his pro-British sympathies stemmed from a much stronger anti-German bias that he shared with the president. House expressed to Wilson his belief that Berlin intended to halt the Monroe Doctrine at the equator to rule South America, primarily Brazil (House, 1914, p. 231). This reinforced the president's own anti-Teutonism. Wilson, the son of a Confederate Presbyterian minister, gave House his opinion that "German philosophy [is] essentially selfish and lacking in spirituality" (House, 1914, p. 149). Yet, House did not trust the Russians either and believed that a defeated Germany would allow Russia to dominate postwar Europe (House, 1914, p. 170).

The same day the First Lady died, the State Department formally requested all belligerents to accept the Declaration of London. Germany and Austria-Hungary quickly complied; France and Russia awaited Britain's decision. On August 7, Grey met with the Dutch ambassador to ask if Holland might join the Entente. Amsterdam refused and declared neutrality (Lambert, 2012, p. 208). On August 10th, Wilson buried his wife in Rome, Georgia. Afterward, he remained secluded within his private railcar and allowed nobody to see him without presidential summons (J. Wilson, 1914). Treasury Secretary McAdoo, the only government official to have access to Wilson, served no official function on the train but accompanied the president as his son-in-law. Wilson admitted to his close friend Mary Hulbert that he had become "dead in heart and body" (Cooper, 2009, p. 263).

In London, intelligence of very large consignments of wheat purchased in New York for delivery to Rotterdam caused a cabinet meeting on August 12th (Lambert, 2012, p. 216). As a result, the government created an interdepartmental subcommittee, called, among other names, the Enemy Supplies Restriction Committee (ESRC). The cabinet charged the ESRC with preventing Rotterdam from acting as a base of supplies for Germany (Lambert, 2012, pp. 219–220).

In London, US Ambassador Walter Hines Page, an old friend of Wilson's, sympathized with the British in the European war. Walter Hines Page had met Woodrow Wilson in 1882. Both were Democrats who developed anti-Republican prejudices during Reconstruction. After election in 1912, Wilson relied heavily upon Page's advice regarding Cabinet appointments before appointing him ambassador to Great Britain (Cooper, 1977, pp. 62–63, 250). The ambassador believed that London ultimately would win the war, and, if Washington did not actively aid Britain, then London might marginalize the United States in the postwar world (Cooper, 1977, p. 288). His tenure as ambassador proved stressful as Page found himself unable to make expenses in London and fell in debt to the amount of $12–15,000 per year (House, 1914, p. 155). Unbeknownst to Page, Wilson broke his solitude on August 17th to arrange a fund to help struggling ambassadors, particularly his friend (House, 1914, p. 142).

That same day, the Admiralty declared: "it is of prime importance to keep the United States of America as a friendly neutral" (Lambert, 2012, p. 237). The next day, Wilson issued a public call that the American people "must be impartial in thought as well as in action" (Wilson, 1914d, p. 394). London followed with the Order in Council of August 20. The British would accept the Declaration of London with their previous reservations becoming modifications to the document. The new Order made no distinction between absolute and conditional contraband destined for neutral ports and imposed a harsh burden of proof upon neutral owners (ASIL, 1915b, p. 459). France and Russia backed Great Britain. The State Department found this Order in Council unsatisfactory. At an August 20th meeting between Secretary of State William Jennings Bryan and Coleville Barclay, British Chargé at Washington, Bryan warned that London's refusal to carry US cargoes to Central Europe made necessary the Ship Registry Law, else US exports would decline (Lambert, 2012, p. 245). Congress passed the bill that day. On August 24th, Wilson sent a bill to Congress to appropriate $30 million to purchase merchantmen to be owned and operated by the US government for five years (Lambert, 2012,

p. 247). London informed Washington that it would not oppose the Ship Registry Bill on condition that German-built ships not carry cargo to German or neutral ports (Lansing, 1914n). The next day, the Admiralty informed Grey that all absolute and conditional contraband bound for Dutch ports would be considered destined for Germany. The orders went out to all flag officers and, within 1 week, 52 neutral ships had been seized (Lambert, 2012, p. 228).

On September 9th, Berlin sent peace feelers to London and Washington (House, 1914, p. 157). Bryan reacted favorably, but Spring-Rice reacted differently, calling the Germans "so unreliable, that their political philosophy [is] so selfish and so unmoral [*sic*] . . ." (House, 1914, p. 163). Spring-Rice hesitated to open peace negotiations, but he also had his doubts about the Americans. He told House that, in his opinion, House himself may have been "the immediate though unconscious cause of this war." He credited House's peacemaking efforts in early 1914 with panicking militarists in Berlin and Vienna. Convinced that Germany and Austria-Hungary had planned to initiate war at some point, Spring-Rice viewed the assassination of Austrian Archduke Franz Ferdinand as merely a pretext for a war. Spring-Rice told House that the German militarists desired war, feared that House's peacemaking efforts made hostilities a "now or never" proposition, and so took advantage of Kaiser Wilhelm II's absence from Berlin to initiate the conflagration (House, 1914, p. 171). With some British government officials holding the Americans in such low esteem, House found negotiations vis-à-vis the Declaration of London difficult. Meanwhile, on September 9th, the ESRC proposed selective harassment and intimidation of neutral shipping. Twelve days later, the Royal Navy captured the *Chester* out of Baton Rouge for Rotterdam (Lambert, 2012, pp. 254–255).

On September 21st, London, in violation of the Declaration, moved certain critical free items to the conditional contraband list, including copper, rubber, and iron ore (Marsden, 1977, p. 492). These items, which had not had any significant military value in 1909, had become militarily useful by 1914. Acting Secretary of State Robert Lansing, filling in for the campaigning Bryan, detailed the history of the Declaration of London in a seven-page cable to Ambassador Page (Hibben, 1929, p. 338). He gave copious legal explanations of why the United States could not accept the August 20th Order in Council and instructed Page to convince Grey to accept the Declaration without amendment (Lansing, 1914a). On September 27th, Lansing proposed to Wilson that the Declaration of London crisis paralleled "the obnoxious Orders in Council of the Napoleonic Wars" (Lansing, 1914k). Lansing reiterated the American position in a cable to Page on September 28th in which Lansing demanded Page see Grey about the matter. The cable which Page received was significantly tamer than the cable that Lansing had at first intended to send, as Lansing had submitted the draft cable to Wilson, who softened Lansing's language for Page and Grey (Lansing, 1914g).

Page cabled back the next day that Grey had refused the American request because the British government had never ratified the Declaration. Grey, however, did not wish to offend the United States (Page, 1914b). That evening, Lansing called upon Spring-Rice to persuade him to influence his government to accept the Declaration, but to no avail (Lansing, 1914h). In conversations on September 28th between Lansing and Spring-Rice, the British ambassador conceded that foodstuffs had become essentially absolute contraband according to the Order in Council. Spring-Rice also intimated that copper, petroleum, and iron from Sweden had become greater worries to London (Lansing, 1914l). House showed Spring-Rice Lansing's original cable meant for Page. The British ambassador expressed his opinion that Lansing's cable was so offensive in its original form that it almost amounted to a US declaration of war against Great Britain (House, 1914, p. 177). Spring-Rice thanked House for overriding the acting secretary of state

(House, 1914, p. 178). Privately, Spring-Rice worried that "the influence of the Germans and especially the German Jews [in the United States] is very great" (Spring-Rice, 1914b). In a private letter, Spring-Rice railed against New York "Jew Bankers . . . very closely connected with the German Government, and seem[ing] to do everything they are told." He further proposed that the Jews were working for world peace and profiting from the war to Germany's benefit (Spring-Rice, 1914c).

By September 29th, intelligence that substantial amounts of refined ores and oils were bound from US ports to Rotterdam and other neutral ports bordering Germany prompted Prime Minister Asquith to consider a plan to mine the eastern North Sea to cut off the Dutch and Belgian ports (Lambert, 2012, p. 260). The next day, the cabinet agreed "in principle" with mining, and Churchill ordered the Naval Staff to draw up mining schedules on October 1st (Lambert, 2012, p. 261).

At the White House, Wilson returned from his wife's funeral. Although technically back at work, Wilson remained in seclusion (Daniels, 1914). In response to a request from John Pierpont Morgan for an audience with the president, Wilson replied that "I find myself so out of spirits that I have for the moment only strength and initiative enough for the absolutely necessary duties of my official day" (Wilson, 1914a). Even more alarming, House found Wilson disengaged from the European war (House, 1914, p. 182). Wilson already suffered from poor health with cerebrovascular disease that predated his election, and his doctor George de Schweinitz found significant hardening of the retinal arteries in March 1914 (Lynn, 2004, pp. 59, 73). The First Lady's death haunted Wilson throughout the Declaration of London episode. At one point, he admitted to House that he was no longer fit to be president and had "no heart" for the position (House, 1914, p. 215). Wilson had even developed a death wish, vouchsafing to House in November that he wished to be assassinated. In other words, without Edith

Axson Wilson, the president wanted to die, a desire he had expressed to House before (House, 1914, p. 227).

On September 30th, Page cabled Lansing with encouraging news. The British government would make a new Order in Council. Grey reported that the Netherlands had agreed to prevent the export of foodstuffs to Germany, allowing Britain to quit confiscating them. Page cabled Lansing: "This whole subject was discussed at a cabinet meeting this morning and it was evident at our conference that the Government desire to meet our wishes so far as the most relentless war ever waged will permit them" (Page, 1914c). Grey's response might have pleased Page, but not Washington. This cable had made plain to Lansing and Wilson that Page was not neutral in thought but instead distinctly sympathetic to Britain and the Entente. Concurrent with the preceding cable, Page forwarded a proclamation from King George V adding certain articles to the conditional contraband list in the Declaration. These articles, formerly on the free list, included copper, lead, glycerin, rubber, and magnetic iron ore (Page, 1914c).

That evening, Wilson and House discussed the crisis. Great Britain would not accept the Declaration as written. The United States, however, would not accept the Declaration amended. House recorded in his diary that Wilson read to him a page from his *History of the American People*:

> The passage said that Madison was compelled to go to war [i.e. the War of 1812] despite the fact that he was a peace-loving man, and desired to do everything in his power to prevent it, but popular feeling made it impossible.
>
> The President said: "Madison and I are the only two Princeton men that have become President. The circumstances of the War of 1812 and now run parallel. I sincerely hope they will not go further." (House, 1914, p. 187)

In this passage of his *magnum opus*, Wilson portrayed the United States as desiring to join in the war on Napoléon Bonaparte but being forced into war with Britain because of British Order in Council restricting American maritime activities. Whether seeded by Lansing or a product of Wilson's own mind, this became Wilson's interpretation of the Declaration episode. House related Wilson's interpretation of history to Spring-Rice, who reported Wilson's 1812 comment, along with the fact that Wilson had written about it to Sir Edward Grey (Spring-Rice, 1914d; House, 1914, p. 187).

An October 4th cable from Grey to Spring-Rice discussed the new Order in Council. Foodstuffs would be removed from the absolute contraband list, but petroleum, copper, rubber, and nickel would replace them (Grey, 1914b). Five days later, the Privy Council superseded the August 20th Order in Council with the draft of a revised Order. The new Order added these items to the absolute contraband list: range finders, hematite iron ore, nickel, ferrochrome, lead, motor vehicles, airplanes, and airships including parts. To the conditional list, the Privy Council added glycerin, fuel oil, and hides (Page, 1914d).

Lansing waited 4 days before replying on October 13th with a curt cable stating that the State Department was considering the draft Order (Lansing, 1914b). The next evening, Lansing met with Wilson to study the draft. That day, Grey warned Spring-Rice that failure to meet agreement with Washington vis-à-vis Rotterdam would leave London either in a dispute with the United States or allow Berlin free access to war material. He noted that copious quantities of petroleum were shipping to Holland and Scandinavian nations bound for Germany (Grey, 1914d).

On October 15th, Lansing submitted to Wilson the draft of a reply to Page listing the American objections to the Order. Lansing wrote Wilson: "The sum total of the objection is that the Order in Council of August 20th is repealed in no particular, but on the contrary, is reenacted with changes and additions which make its provisions even more objectionable" (Lansing, 1914i). Meanwhile, Page had not heard any word from Washington since the cable of October 13th. The rude tone of the cable made him fear that Washington would reject the new Order in Council and thus exacerbate the crisis. That day, Spring-Rice noted to Wilson that US exports of all kinds had increased due to the war. In so doing, Spring-Rice implied not only that the new Order in Council would not hurt US trade as much as claimed but also that US exports helped Germany create a 2-year surplus before the war's beginning. He then asked Wilson to acquiesce in British enforcement of the blockade vis-à-vis Rotterdam, as the US had in times past invoked the same doctrine vis-à-vis Mexico (Spring-Rice, 1914a). Page warned Bryan that London was willing to risk war with the United States rather than allow free neutral trade with the Netherlands or agree to the 1909 version of the Declaration of London (Page, 1914i).

In desperation, Page sent two letters to Wilson on October 15th; one he telegraphed, the other he sent by steamer. In the cable, Page said:

> For the President: Present controversy about shipping. I cannot help fearing we are getting into deep water needlessly. . . . The present controversy seems here, close to the struggle, academic . . . The question seems wholly different here from what it probably seems in Washington. There it is a more or less academic discussion. Here it is a matter of life and death for English-speaking civilization. (Page, 1914e)

In his letter via steamer, Page wrote more candidly: "This resolution reads here like a Sunday-school resolution passed in Kansas requesting cruel Vesuvius to cease its eruption, which destroys villages of innocent people"

(Hendrick, 1923, p. 164). With all the diplomatic and financial strains in his life, Page's ulcer resumed. Insomnia also became frequent (Cooper, 1977, p. 293).

The next day, October 16th at 1300 hours, Page received a cable from Washington. It was not from Wilson, but Lansing. Lansing ordered Page to convince Grey to accept the Declaration without amendment and to obtain the retraction of all Orders in Council altering the Declaration. As a final insult, Lansing ordered Page to present these demands as personal requests for which Washington would not have to bear responsibility (Lansing, 1914c). Wilson sent Page no response to his desperate appeals of the prior evening. Two hours later, Page received another cable from Washington, again from Lansing. This cable detailed the US objections to the Order in Council and again pressed for British acceptance of the unadulterated Declaration (Lansing, 1914c).

Finally, at 1800 hours, Page received the cable for which he had waited. Unfortunately, Wilson's response was not what Page had hoped: "Your October 15, 11 p.m. Beg that you will not regard the position of this government as merely academic. Contact with opinion on this side of the water would materially alter your view." Wilson ended with: "This is private and for your guidance" (Wilson, 1914b). Wilson had dismissed Page's reasons apparently without a thought. Having had Wilson abandon him and Lansing ignore him, Page fell into a deep depression (Gregory, 1970, p. 70).

On October 19th, Page reported to Lansing on his meeting with Grey. Grey again had refused to accept the Declaration of London unamended. As to the Order in Council, Grey told Page that the British government would issue a statement rescinding both Orders and would accept the Declaration with two reservations. They would add to the contraband list those items in the Order in Council to which Lansing did not object. The

Royal Navy would reserve the right, however, to stop contraband cargoes which, although bound for neutral ports, ultimately would find their way to a belligerent. Grey requested that the United States not protest the British position but instead challenge the stopping of individual ships on a case by case basis (Page, 1914f). Page informed Lansing: "This finally ends all hope of his acceptance of the declaration entire. He is courteous, appreciative, and willing to go any length he can to meet us, but he will not accept the declaration for the reasons given" (Page, 1914f).

Page's cable arrived at Washington on October 20th at 0800 hours. That same day, Spring-Rice handed Lansing two cables which he had received from Grey. In them, Grey ordered Spring-Rice to try to convince the Americans to accept the deal he had outlined to Page earlier that day (Grey, 1914a). By handing over the cables to Lansing, Spring-Rice had produced the desired reaction. Lansing wrote the president that, due to the October 19th cable from Page and the two cables handed to him by Spring-Rice, he had decided to drop the Declaration of London: "It seems to me that in view of the rigid attitude of the British Government further attempts to obtain an agreement on the Declaration of London would be useless. We must, therefore, stand on the rules of international law which have been generally accepted without regard to the Declaration" (Lansing, 1914j). Walter Hines Page had won, but nobody informed him of the decision. That same day, Page cabled Lansing and Wilson begging for acceptance of the British position. At the end of the cable, Page warned Lansing against further pursuing the Declaration: "The hope of every patriotic man on either side of the ocean will be disappointed and such good will as is now left in the world will be gone" (Page, 1914a). Page desperately needed an answer from Washington. On October 21st, Lansing sent Wilson a memo containing the draft of the notice to Page that the United States had dropped the Declaration. Wilson responded to Lansing, "Approved as altered. Would it not be well to show this to S[pring]-R[ice] before you send it, - for any comment or suggestion he might have to make?" (Lansing, 1914f). Effectively, Wilson had informed the British ambassador about a significant alteration in US foreign policy before he informed the American ambassador.

By October 21st, Page grew disturbed. Not only did the US position not admit a diplomatic solution, but Page also found Lansing's proposal to have him present US plans as unofficial and unendorsed to be insulting if not duplicitous. Page complained to Wilson that Lansing was "argumentative," "critical," and possessed a "tone of disgust" (Page, 1914h). Page also cabled Lansing on October 21st: "Great Britain is within its rights in adopting now the rules which it proposes to adhere to" (Page, 1914a). The same day, Page sent a cable to his son Arthur that he meant for his son to show House. In the cable, Page vented his frustrations against Lansing, the man who had ignored him so often. "God deliver us, or can you deliver us, from library lawyers," wrote Page, referring disparagingly to Lansing's legal background (Seymour, 1926, p. 305). House presented Page's cable to the president along with the remark, "I hardly know to what he refers, but perhaps you do. It may be the Declaration of London matter" (Seymour, 1926, p. 306). House already had forgotten about Page's concern. If it was unfair for House to dismiss Page so quickly, considering Washington had not yet informed Page of the end of the crisis, then Wilson's ensuing comments were just as unfair.

The cable infuriated fellow former lawyer Wilson, yet neither Wilson nor Lansing understood that the United States could make major concessions to Great Britain regarding the Declaration and nevertheless remain neutral according to international law (Gregory, 1970, p. 74). Wilson told House that Page's attitude disturbed him. Wilson insisted that, as ambassador, Page needed to see issues in the same light as seen in the United States. In other words, Page needed to see things Wilson's way. Wilson also stated that Page's sympathy for the British case might prove a danger (Seymour, 1926, p.

306). Wilson, through House, warned Page to curb his Anglophilia (House, 1914, p. 239).

On October 22nd, Page sent a letter to Washington, by steamer, threatening to resign as ambassador if the United States did not accept the British proposals (Gregory, 1970, p. 75). In a short cable at 1600 hours, Lansing informed Page that the United States had backed down (Lansing, 1914d). In Page's response the next day, he related that "withdrawal of the Declaration of London has most admirable effect all around" (Page, 1914g). On October 24th, Grey notes to Spring-Rice with satisfaction that the United States had dropped the Declaration in its entirety and that Britain would detain all ships carrying petroleum and copper until verification that they would dock and unload their cargoes for consumption in neutral countries (Grey, 1914c). Finally, Whitehall issued an Order in Council on October 29th, 1914, which in large measure granted Washington's interpretation of non-inference vis-à-vis conditional contraband to neutral ports (ASIL, 1915b, p. 459).

Regarding the Declaration of London, Woodrow Wilson informed Page on November 10th, "Any mistakes that were made then can now be easily forgotten" (Wilson, 1914c). Page could not forget, as his health problems started because of the crisis but did not end with it; rather, they continued for the rest of his life (Gregory, 1970, p. 70). Wilson's mental state also continued to deteriorate. In a White House meeting with a delegation led by William Monroe Trotter, Wilson lost his temper in a discussion about segregation of federal agencies (Cooper, 2009, p. 270). Also, the world continued to deteriorate around Page and Wilson. British violations of the unratified Declaration of London led in part to Berlin's February 4th, 1915, declaration of unrestricted submarine warfare. Germany made clear that its U-boats would target not only hostile merchantmen in British waters but also British merchantmen flying neutral flags (Marsden, 1977, p. 501).

Conclusion

In the end, Wilson backed down and a US–British front in the Great War was avoided. Yet President Wilson jeopardized US neutrality and nearly provoked a war with Great Britain in his attempt to redefine international neutrality law. This resulted in part from the biases of the principals involved. Wilson was culturally Anglophilic, but his partiality for England did not extend to entangling the United States in its wars. House was also an Anglophile, but the two men's actions during the Declaration of London crisis might better be characterized as Teutonophobic neutrality: both hated Germany more than they favored Britain. The most Anglophilic official, Walter Hines Page, ironically found his council increasingly marginalized as biased throughout the crisis. On the British side, Spring-Rice's anti-Semitism led him to interpret American neutrality as part of a German Jewish plot against London.

Wilson's mental health also played a role in the crisis. Already in poor health for years, and suffering hardening of the retinal arteries, his stubbornness during the crisis, particularly vis-à-vis his friend Page, could reflect damage from his multiple strokes (Lynn, 2004, p. 59). Then after Ellen's death, Wilson withdrew from fully engaging with the world and his presidency for the rest of 1914. He focused only tangentially upon the European war and only somewhat half-heartedly upon the Declaration of London. He expended most of his efforts upon the Ship Registry Bill and the succeeding Purchase Bill, both of which served the interests of American business and endangered American neutrality.

This could indicate what Wilson considered his most important job as president: the expansion of American business. Neutrality served American business interests, but insisting upon acceptance of the Declaration of London endangered that neutrality, so Wilson eventually dropped it.

Unfortunately, in part due to being disengaged for whatever reasons, it took the president months before he finally accepted the advice of Ambassador Page and ceded the issue to the British. In the meantime, Wilson nearly found himself in league with a culture he despised and at war against a culture he respected.

References

American Society of International Law. (1915a). Status of the Declaration of London. *The American Journal of International Law*, *9*, 199–202.

American Society of International Law. (1915b). Seizure and Detention of Neutral Cargoes – Visit and Search – Continuous Voyage. *The American Journal of International Law*, *9*, 456–461.

Bryan, W.J. to J. MacMurray, 7 August 1914. (1914). In *Foreign Relations of the United States* [hereafter *FRUS*] *1914: Supplement, The World War*. Washington: Government Printing Office, 1928.

Cooper, J.M. (1977). *Walter Hines Page: The Southerner as American 1855-1918*. Chapel Hill: University of North Carolina Press.

Cooper, J.M. (2009). *Woodrow Wilson: A Biography*. New York: Vintage Books.

Daniels. J. to Bryan, W.J., 12 August 1914. (1914). in *FRUS 1914: Supplement, The World War*. Washington: Government Printing Office, 1928.

Gregory, R. (1970). *Walter Hines Page: Ambassador to the Court of St. James's*. Lexington KY: University Press of Kentucky.

Grey to Spring-Rice, 17 and 19 October 1914. (1914a). in *FRUS 1914: Supplement, The World War*. Washington: Government Printing Office, 1928.

Grey to Spring-Rice, 4 October 1914. (1914b). in *The Papers of Woodrow*

Wilson, vol. 31: September 6 – December 31, 1914. Princeton: Princeton University Press.

Grey to Spring-Rice, 24 October 1914. (1914c). in *The Papers of Woodrow Wilson, vol. 31: September 6 – December 31, 1914*. Princeton: Princeton University Press.

Grey to Spring-Rice, 14 October 1914. (1914d). in *The Papers of Woodrow Wilson, vol. 31: September 6 – December 31, 1914*. Princeton: Princeton University Press.

Gould, L. (2008). *Four Hats in the Ring: The 1912 Election and the Birth of Modern American Politics*. Lawrence KS: University Press of Kansas.

Gould, L. (2014). *Chief Executive to Chief Justice*. Lawrence KS: University Press of Kansas.

Gould, L. (2016). *The First Modern Clash Over Federal Power: Wilson versus Hughes in the Presidential Election of 1916*. Lawrence KS: University Press of Kansas.

Hendrick, B. (1923). *The Life and Letters of Walter H. Page, vol. 1*. Garden City: Doubleday, Page & Co.

Hibben, P. (1929). *The Peerless Leader: William Jennings Bryan*. New York: Farrar and Rinehart.

House, Edward M. (1914). *Diary*. MS 466, Edward Mandell House Papers, Series II, Diaries, Volume 2. New Haven CT: Yale University Library.

Lambert, N. (2012). *Planning Armageddon: British Economic Warfare and the First World War*. Cambridge: Harvard University Press.

Lansing to Page, 26 September 1914. (1914a). in *FRUS 1914: Supplement, The World War*. Washington: Government Printing Office, 1928.

Lansing to Page, 13 October 1914. (1914b). in *FRUS 1914: Supplement, The World War*. Washington: Government Printing Office, 1928.

Lansing to Page, 16 October 1914. (1914c). in *FRUS 1914: Supplement, The World War*. Washington: Government Printing Office, 1928.

Lansing to Page, 22 October 1914. (1914d). in *FRUS 1914: Supplement, The World War*. Washington: Government Printing Office, 1928.

Lansing, R. "Memorandum: Course to be Pursued to Preserve the Status Quo in China", 7 August 1914. (1914e). in *FRUS: The Lansing Papers, 1914-1920, vol. 1*. Washington: Government Printing Office, 1939.

Lansing to Wilson, 21 October 1914. (1914f). in *FRUS: The Lansing Papers, 1914-1920, vol. 1*. Washington: Government Printing Office, 1939.

Lansing, R. "Draft Telegram", 28 September 1914. (1914g). in *FRUS: The Lansing Papers, 1914-1920, vol. 1*. Washington: Government Printing Office, 1939.

Lansing, R. "Memorandum by the Acting Secretary of State", 29 September 1914. (1914h). in *FRUS: The Lansing Papers, 1914-1920, vol. 1*. Washington: Government Printing Office, 1939.

Lansing to Wilson, 15 October 1914. (1914i). in *FRUS: The Lansing Papers, 1914-1920, vol. 1*. Washington: Government Printing Office, 1939.

Lansing to Wilson, 20 October 1914. (1914j). in *FRUS: The Lansing Papers, 1914-1920, vol. 1*. Washington: Government Printing Office, 1939.

Lansing to Wilson, 27 September 1914. (1914k). in *The Papers of Woodrow Wilson, vol. 31: September 6 – December 31, 1914*. Princeton: Princeton University Press.

Lansing, R. "Memorandum of Conversation with British Ambassador", 29 September 1914. (1914l). in *The Papers of Woodrow Wilson, vol. 31: September 6 – December 31, 1914*. Princeton: Princeton University Press.

Lansing to Wilson, 19 October 1914. (1914m). in *The Papers of Woodrow Wilson, vol. 31: September 6 – December 31, 1914*. Princeton: Princeton University Press.

Lansing to Wilson, 24 August 1914. (1914n). in *The Papers of Woodrow Wilson, vol. 30: May 6 – September 5, 1914*. Princeton: Princeton

University Press.

Link, A. (1960). *Wilson: The Struggle for Neutrality, 1914-1915*. Princeton: Princeton Univ. Press.

Lynn, K. (2004). The Hidden Agony of Woodrow Wilson. *Wilson Quarterly, 28*, 59-92.

Marsden, A. (1977). The Blockade. in *British Foreign Policy under Sir Edward Grey*. Cambridge: Cambridge University Press. Pp. 488-517.

Martin, C. (2009). The Declaration of London: A Matter of Operational Capability. *Historical Research, 82*, 731-755.

New York Times, "Put World Trade Under Flag of U.S., President's Plan", 1 August 1914. (1914). in *The Papers of Woodrow Wilson, vol. 30: May 6 – September 5, 1914*. Princeton: Princeton University Press.

Page to Lansing, 21 October 1914. (1914a). in *FRUS 1914: Supplement, The World War*. Washington: Government Printing Office, 1928.

Page to Lansing, 29 October 1914. (1914b). in *FRUS 1914: Supplement, The World War*. Washington: Government Printing Office, 1928.

Page to Lansing, 30 October 1914. (1914c). in *FRUS 1914: Supplement, The World War*. Washington: Government Printing Office, 1928.

Page to Lansing, 9 October 1914. (1914d). in *FRUS 1914: Supplement, The World War*. Washington: Government Printing Office, 1928.

Page to Lansing, 15 October 1914. (1914e). in *FRUS 1914: Supplement, The World War*. Washington: Government Printing Office, 1928.

Page to Lansing, 19 October 1914. (1914f). in *FRUS 1914: Supplement, The World War*. Washington: Government Printing Office. 1928.

Page to Lansing, 23 October 1914. (1914g). in *FRUS 1914: Supplement, The World War*. Washington: Government Printing Office, 1928.

Page to Wilson, 21 October 1914. (1914h). in *The Papers of Woodrow Wilson, vol. 31: September 6 – December 31, 1914*. Princeton: Princeton University Press.

Page to Bryan, 15 October 1914. (1914i). in *The Papers of Woodrow Wilson, vol. 31: September 6 – December 31, 1914.* Princeton: Princeton University Press.

Ross, G. (1970). *Walter Hines Page: Ambassador to the Court of St. James's.* Lexington: Univ. Press of Kentucky.

Ross, G. (1971). *The Origins of American Intervention in the First World War.* New York: W.W. Norton.

Savage, C. (1936). *Policy of the United States toward Maritime Commerce in War, vol.2: 1914-1918.* Washington DC: US Department of State.

Seymour, C. (1926). *The Intimate Papers of Colonel House.* Boston: Houghton Mifflin Co.

Schuler, F. (2011). *Secret Wars and Secret Policies in the Americas, 1842-1829.* Albuquerque: University of New Mexico Press.

Spring-Rice to Wilson, 15 October 1914. (1914a). in *The Papers of Woodrow Wilson, vol. 31: September 6 – December 31, 1914.* Princeton: Princeton University Press.

Spring-Rice to Grey, 25 August 1914. (1914b). *The Letters and Friendships of Sir Cecil Spring-Rice, vol. 2.* London: Constable & Co., Ltd.

Spring-Rice to Chirol, V., 27 November 1914. (1914c). *The Letters and Friendships of Sir Cecil Spring-Rice, vol. 2.* London: Constable & Co., Ltd.

Spring-Rice to Grey, 1 October 1914. (1914d). *The Letters and Friendships of Sir Cecil Spring-Rice, vol. 2.* London: Constable & Co., Ltd.

Wilson, J. to Wilson, K. and A., 11 August 1914. (1914). in *The Papers of Woodrow Wilson, vol. 30: May 6 – September 5, 1914.* Princeton: Princeton University Press.

Wilson to Morgan, 13 August 1914. (1914a). in *The Papers of Woodrow Wilson, vol. 30: May 6 – September 5, 1914.* Princeton: Princeton University Press.

Wilson to Page, 16 October 1914. (1914b). in *FRUS 1914: Supplement, The World War*. Washington: Government Printing Office, 1928.

Wilson to Page, 10 November 1914. (1914c). in *The Papers of Woodrow Wilson, vol. 31: September 6 – December 31, 1914*. Princeton: Princeton University Press.

Wilson, W., "An Appeal to the American People", 18 August 1914. (1914d). in *The Papers of Woodrow Wilson, vol. 30: May 6 – September 5, 1914*. Princeton: Princeton University Press.

2

Xenophobia Unleashed: Anti-German Attitudes and Policies During WWI

B.D. Mowell
American Military University

Abstract

The rise in anti-German sentiment associated with WWI provides an excellent case study in wartime xenophobia and nativism and illustrates how the latter can manifest even in democratic societies. Largely baseless perceptions of the German American community as disloyal became more widespread as US sentiment shifted away from neutrality and reached a state of near hysteria in many areas upon US entry into the war in 1917, though such xenophobic reactions generally did not target ethnicities associated with Germany's allies. While overt violence directed against German Americans was uncommon, other forms of maltreatment were widespread including ostracization, employment discrimination, and boycotts of German American-owned businesses and cultural vestiges. Anti-German xenophobia also marked one of the few occasions in modern US history in which discriminatory and nativist public sentiment became widely codified into local, state and even federal policies leading to actions such as widespread bans of German-language instruction by many school districts across the country. Although some 2,000 German-born aliens were ultimately interned during the war, calls for large-scale internment or deportation of German Americans went unheeded largely because of their sheer numbers and political influence. The treatment of German Americans during WWI illustrates the potential dangers of xenophobic public sentiment and "extra-Constitutional" government actions tolerated under the guise of wartime security measures.

Keywords: Alien and Sedition Acts, discrimination, German Americans, nationalism, nativism, xenophobia

Xenophobia Unleashed: Anti-German Attitudes and Policies During WWI

As the result of World War I, a range of xenophobic actions were directed against persons of German ancestry or vestiges of German culture. Many such actions were symbolic and inane efforts to strike a public blow against German culture or heritage, such as initiatives to rename sauerkraut, German measles, and German Shepherd dogs *liberty cabbage*, *liberty measles*, and *Alsatian Shepherds* respectively (Kazal, 2004, p. 176). However, many manifestations of nativism and xenophobia directed against those of German descent were more reprehensible in nature. While violence directed toward German-Americans was not widespread, it did occur. Approximately 30 victims are thought to have been killed by hyper-nationalist vigilantes, the most infamous case being that of immigrant Robert Prager who was lynched in Illinois by a large mob after having been falsely accused of plotting to sabotage the mine in which he worked (Kirschbaum, 2015). More commonly, hundreds of people of German descent were beaten or tarred and feathered, thousands dismissed from employment, and countless numbers subjected to verbal abuse, threats, and harassment.

Oddly, a double standard existed wherein xenophobic policies and public hysteria were largely specific to all things German and generally not also projected onto foreign nationals from nations allied with Germany, seemingly delegitimizing any argument that anti-German actions and policies were to some degree justified by domestic security concerns. While considerably smaller in size than the German population in the country, the numbers of immigrants from Austria-Hungary, Bulgaria, and the Ottoman Empire residing in the United States were substantial. Census data from 1910 indicates among foreign-born persons in the United States, 626,341 reported having been born in Austria, 11,498 in Bulgaria, 495,609 in Hungary, and 59,729 in Turkey—a combined total of 1,193,177 from other Central Powers nations, compared to 2,311,237 who indicated Germany as their birthplace (US Census Bureau, 1999). Many in the United States advocated going to war against all of the Central Powers, perhaps most famously Theodore Roosevelt who argued that "there is no use in making four bites of a cherry" or "going to war a little, but not much," but public enmity and domestic security concerns remained focused almost entirely upon Germany both leading up to and during the war (Thompson, 2014, p. 215). For example, although wartime immigration restrictions were implemented to prevent their entry into the United States, no systematic regulation of foreign nationals having already emigrated from Austria-Hungary, Bulgaria, or the Ottoman Empire was undertaken, allowing nationals from the latter countries unfettered movement, which many utilized to pursue employment in ammunition plants or in other sensitive capacities supporting the war effort, a flagrant double standard which did not go unnoticed or without criticism in the press at the time (e.g., *The New York Times*, 1917).

This chapter provides an overview of xenophobic policies and actions directed against German culture and persons of German descent in the United States leading up to and during WWI. While other episodes of unchecked nativism and maltreatment of ethnicities exist in US history, the xenophobia directed against German people and culture remains one of the least understood. The chapter explores the political and societal circumstances contributing to an emerging rhetoric of questioned loyalty and how this concomitantly evolved into unjustified public hysteria and institutionalized practices, including codified discrimination and even internment under the guise of wartime national security.

German-Americans: Historical Background

While relatively small numbers of German-speaking immigrants came to the United States in the 17th and 18th centuries, large-scale German immigration began in the mid–late 19th century, at which point they became the first non-English speakers to enter the country in substantial numbers. Germans emigrated from their homeland for various reasons, including economic opportunity or political and religious freedom, and of the approximately seven to eight million who left their ancestral homeland between the early 1800s and WWI, around 90% settled in the United States (Library of Congress, 2017). The 1880s marked the peak period of German immigration, with around 1.5 million new arrivals, but, by the turn of the century as the German economy expanded and more jobs were created in the newly formed Empire, levels of immigration had declined to approximately 300,000 arrivals in the first decade of the 20th century. German-Americans proved themselves to be a productive element of US society and played vital roles in the country's early growth, establishing farms and serving as urban laborers—frequently in skilled trades and often founding businesses, many of which became major companies and continue to bear the names of their German-American founders today, including Anheuser-Busch, Boeing, Heinz, Levi-Strauss, Merck, Pabst, Pfizer, and Steinway, for example.

Although some immigrants remained monolingual, to a great degree German immigrants embraced an American cultural identity, including learning English. Yet like many European immigrant communities, many German-American enclaves retained a sense of cultural distinctiveness and community pride and strove to keep traditions alive, including their language. While English would customarily be spoken at work and many public places, German often remained the primary language in the home environment, predominantly German-American churches, social clubs, and certain other venues. Areas with large enough German-American communities also had German-language businesses and even newspapers—around 1,000 of which were being published by 1890, accounting for some 80% of foreign-language press published in the United States (Grohsgal, 2014). The influence of German-American communities, especially in many Midwestern and northern states, often facilitated the inclusion of German-language instruction in many US schools and colleges, and in many German-American enclaves, it was the primary language of instruction for immigrant children, though the latter was more common in private schools established within the communities. Though even larger-scale anti-German backlash would manifest in a few decades, in the late 1880s nativist and xenophobic movements emerged in the United States seeking to ban the use of foreign languages and even eliminate instruction in languages other than English, efforts that often targeted German speakers (Petty, 2013).

By the early 1900s, the German-American community had become very significant demographically. Approximately 9% of the US population were either immigrants who had been born in Germany or the children or grandchildren of German immigrants, making German-Americans the largest ethnic group in the country at the outset of the 20th century (German Historical Institute, 2017a). As of 1910, German immigrants were the largest single ethnic group in 18 states as well as statistically in the United States overall, accounting for nearly 20% of all foreign-born persons in the country (Krogstad & Keegan, 2015). While Americans of German ancestry could be found throughout the country, they were largest in number and had the most significant impacts upon the demographic and human landscapes of the upper Midwestern states, including Wisconsin, Minnesota, Iowa, Illinois, and Nebraska. It should be noted that German-Americans were not a homogeneous, monolithic group, as they had come from culturally diverse regions and backgrounds and did not all share the same political views, experiences, or culture traits. Religious diversity among their ranks exemplifies the latter, as most German immigrants from northern areas tended to be Protestants, most from the southern areas were Catholic, and a sizable minority of arrivals to the United States were German Jews.

Increasing Rhetoric and Questioned Loyalties

With the establishment of a unified Germany and its subsequent global ambitions as an emerging economic and geopolitical power, the German government eventually sought to yoke the strength of the international German diaspora, the largest segment of which resided in the United States. By the turn of the century, efforts were undertaken by Berlin to coordinate with the German diaspora in promoting the teaching of German language and culture and to use the diaspora communities to promote a favorable view of Germany and its policies internationally. Given the diversity of German-Americans in terms of political views and otherwise, the mobilization of the diaspora community behind a new trans-German nationalism was never truly realized (Manz, 2014). However, as Germany found itself increasingly at odds with European colonial competitors in the early 20th century, and certainly with the outbreak of WWI in 1914, many members of the German-American community as well as the German-language press were publicly supportive of their ancestral homeland and often worked to influence a favorable public opinion of Germany in the neutral United States. For example, some German-Americans were alienated by discrimination against their language and culture and were threatened by their declining numbers and political clout in many cities, stemming from decades of decreased German immigration coupled with increasing arrivals of other ethnicities. As a result, a minority of German-Americans—particularly in larger cities in the north and Midwest—identified at least to some degree with the emergence of a German Empire and resisted the "Americanization" efforts targeting immigrants and their descendants; some even embraced notions of inherent German superiority being floated in the motherland (Dobbert, 1967; O'Connor, 1968). The latter was also being driven in part by broader trends of ethnic-nationalism which resurged in the late 19th and early 20th centuries in the United States and much of the western world and was of course not specific to persons of German ancestry.

While most German-American publications advocated neutrality and did not spout propaganda, some did in fact function as apologists for Germany's role in the war, casting blame for the conflict upon Russia's territorial ambitions, France's desire for revenge at losing the Franco-Prussian war, or Britain's war profiteering (Franck, 1989; Rippley, 1976). Early in the European conflict, a prominent German-American newspaper characterized the Kaiser as a "paragon of statesmanship," arguing that he had done everything within his power to mediate peace between Austria and Russia (Doenecke, 2011, p. 26). Such views sharply contrasted with those expressed in the mainstream US media which had been generally sympathetic to the allied cause from the war's inception, thus partly incentivizing the

German-American publications that were sometimes printed in English to reach a broader audience and present the US public with what they regarded as a valid alternative perspective (Franck, 1989). Rather than steering public opinion toward the German perspective—or even favoring neutrality—such efforts tended to arouse suspicion among many in the US population and led to increased public scrutiny and debate concerning the degree to which the German-American community could be trusted.

Following US entry into the war in 1917, subscriptions to German-language newspapers in the United States plummeted, and in October of that year, Congress enacted legislation to regulate the foreign-language press, the first such regulation of the press in national history (Rippley, 1976). The new law required any content related to the war to be submitted to the local postmaster for translation, review, and censoring until the government was satisfied as to the publication's loyalty, at which point a permit could be issued exempting the publication from further official scrutiny. However, to a great degree, the regulatory efforts became a moot point as most of the publications had ceased operation by 1918 and many that remained had transitioned to publishing in English (Rippley, 1976). Whereas nearly 900 German-language publications existed in the United States in 1894 at the zenith of the phenomenon, only 172 were known to be in existence in 1930—most of which were small-circulation newsletters of German-American civic or religious organizations, a testament to the long-term impact of wartime xenophobia (Rippley, 1976, p. 164–166). Also, facing intense wartime public scrutiny, many German-American organizations across the country disbanded, including many chapters of the once influential German-American alliance (Rippley, 1976).

As WWI dragged on in Europe, German-Americans were sharply divided in their political opinions. A minority actively supported Germany's position and perhaps even hoped that circumstances might eventually allow their adopted homeland to enter the war on Germany's side. More commonly and more realistically, a position of neutrality was advocated, mirroring the prevailing view of most Americans that the country should stay out of the war. In the opening years of the war, German-American rallies were commonplace and well-attended in many parts of the United States. Often large crowds extolled pride in the achievements of German civilization and loudly and publicly proclaimed admiration for the German Empire, though as the war continued and the possibility of direct US involvement increased, such outward displays of ethnic pride diminished and eventually ceased entirely (Kamphoefner, Helbich, & Sommer, 1991). However, many German-Americans, either due to their immigrant ancestors' cultural assimilation in the United States and thus having no sense of kinship to Germany or alternatively due to the persecution they or their ancestors (i.e., German Jews) may have experienced prior to emigrating, did not identify with Germany and harbored no sympathies for its wartime struggles (O'Connor, 1968, p. 265). Also, smaller numbers of German-Americans or their ancestors had immigrated to the United States prior to the initial creation of the unified German state in 1871, and, having ancestral ties to countries which no longer existed—such as Bavaria, Prussia, or Saxony—they had less sense of a connection to the relatively new Imperial German state.

Interestingly, the US temperance crusade was in part a nativist and xenophobic movement, and it has been suggested that the efforts supporting Prohibition may have also factored into public distrust and animosity being directed toward German-Americans (e.g. Davis, 2014; Riddley, 1976). While not all German-Americans imbibed, beer consumption in the United States was in part popularized by German immigrants, and by the turn of the 20th century, German-American owned breweries, beer gardens and saloons, and German styles of beer such as lager had become commonplace, as had stereotypes of Germans as drinkers. Importantly, German-American publications, organizations, and other activities were often financially supported by breweries and other alcohol-related businesses which sought to mobilize public opinion against Prohibition. By 1908, Prohibition supporters, characterizing German-Americans as immoral and culturally alien nonconformists, pursued efforts to restrict the voting rights of German immigrants particularly via heightened reading and language requirements, hoping this might help tip the balance in favor of an alcohol ban (Riddley, 1976). Although the latter endeavors were unsuccessful, the efforts of the temperance movement to associate German-Americans with moral turpitude and as having a negative foreign influence upon US society likely helped pave the foundation for other negative public perceptions and war-era suspicions (Davis, 2014; Fiebig-von Hase & Lehmkuhl, 1997).

Lusitania as a Tipping Point in Anti-German Xenophobia

As a rising tide of public opinion in the United States had already begun to reflect anti-German sentiment and distrust of the German-American community, the sinking of the British ocean liner RMS *Lusitania* by a German submarine in 1915 was a tipping point both in terms of xenophobia and in beginning to sway many former advocates of US neutrality in favor of entering the war against Germany and its allies. German submarine warfare in the North Atlantic had been progressively intensifying in response to the British blockade and in an effort to reduce the maritime flow of supplies to Britain and its European allies. Early in WWI, partly due to increasing British use of hidden deck guns aboard commercial ships, Germany implemented a policy of unrestricted submarine warfare wherein ships were attacked without warning. On May 7th, the *Lusitania*, which had originated its voyage in New York, was torpedoed off the coast of Ireland and quickly sank, killing 1,198 people, including 124 Americans (Eardley, 2014). The German government argued the attack was justified as, in addition to passengers, they alleged that the ship was carrying large amounts of ammunition and other war material and had sailed through a designated war zone, failing to warn passengers not to seek passage on the British-flagged ship as it was deemed a wartime target for search or attack.

However, public outrage toward Germany in the United States and abroad was ignited by the sinking and subsequent sensationalized accounts of the attack, which often included photos of recovered bodies of victims. Anti-German riots erupted in Britain in which angry, and often inebriated, mobs attacked persons of German heritage and vandalized their businesses and homes (Gullace, 2005). While riots did not erupt in US cities in 1915, anti-German attitudes became more widespread and more acceptable to share openly; indeed, some scholars (e.g. Trommler, 2009) cite the sinking of the *Lusitania* as the single-most important catalyst for the shift in US sentiment toward entry into the war. The sinking became a potent propaganda tool for interventionists as a symbol of evil, malice, and aggression and as a cowardly act that warranted retaliation (Trommler, 2009). The outcry over the *Lusitania* also enabled the arguably already pro-British position of the Woodrow Wilson administration to shift further from strict neutrality via the increased provision of loans and other aid and increased export of supplies—including ammunition to London—with decreasing political opposition in the United States, though many German-American organizations continued to lobby for neutrality (O'Connor, 1968). The acquiescence of the German government to US demands of compensation for *Lusitania* victims and a temporary moratorium on unrestricted submarine warfare averted US entry into the war in 1915. However, the tide had inexorably turned concerning US public opinion, with the majority at least favoring biased neutrality, which permitted aid to Britain and France, and larger numbers than ever supported US entry into the war.

Those in the United States who favored providing at least indirect assistance to Britain and the Allies regarded themselves as more American and more patriotic than those who advocated strict neutrality, whom such Americans viewed as some form of "others" (O'Connor, 1968). Such attitudes added fuel to the fire of the "hyphenated American" political rhetoric, wherein many nationalists believed segments of US society held divided loyalties. Such attitudes were perhaps best disseminated on the national stage by former President Theodore Roosevelt, who argued that the nation could not endure "half hyphenated and half American" (Thompson, 2014, p. 125). A few months after the sinking of the *Lusitania*, Roosevelt stressed in a fiery speech that "there is no room in the US for hyphenated Americans" and that "we must unsparingly condemn any man who holds any other allegiance" (Manning, 2014, p. 16). Increasingly within national discourse, xenophobic nationalists called for "100% Americanism," wherein immigrant groups or any other community with a distinct ethnic or cultural heritage—particularly German-Americans—should not retain multiple identities that may run afoul of what nationalists regarded as correct ways of thinking (Trommler, 1998, p. 32–33).

German-Americans and others who continued to oppose US involvement in the war came under increasingly intense public criticism and rebuke. Earlier proposals that, in order to preserve strict US neutrality, a moratorium on loans and an embargo on exports in general, or at least on military-related supplies in particular, be imposed upon all warring nations of Europe were effectively dead. Increasingly, polities inside the United States, including progressives and other leftists, pacifists, and such German-American groups as the National German-American Alliance that supported continued neutrality, were viewed with suspicion. An example of such paranoia can be seen in a widely read 1915 book, titled *German Conspiracies in America*, which characterized all German-Americans as un-American and incapable of assimilation, and it alleged, among other things, that their communities were riddled with spies and saboteurs ready to act on behalf of the kaiser (O'Connor, 1968). Trommler states that in the 2-year interval between the *Lusitania* and US entry into the war, German-Americans presented the largest target for nativist public agitation and came to be identified as "different, dissenting, disagreeable or unpatriotic" (Trommler, 2009, p. 247). Such vitriol and suspicion were in general not publicly directed at other pro-neutrality ethnic groups, including Irish- or Scandinavian-Americans. When war was finally declared in April 1917, the past political rhetoric of formerly pro-neutrality groups was often scrutinized as evidence of an unpatriotic or traitorous "enemy within," even when former supporters of neutrality—including many German-Americans—fulfilled their civic duty by enlisting in the military, buying war bonds, or otherwise supported the national war effort (Trommler, 2009, p. 248).

Enactment of Legislation

US entry into the war in April 1917 facilitated the crafting and implementation of new federal legislation based upon the precedents established in the Alien and Sedition Acts of 1798. Under the terms of the latter, foreign nationals were deemed a security risk, and those who held citizenship in a nation hostile to the United States were subject to imprisonment or deportation, as was also the case for those engaging in disloyal speech. Sizable public protests were common in many US cities, with large numbers of Americans objecting to either entry into the war in general or the implementation of a national draft via the 1917 Selective Service Act to bolster the diminutive size of the US military in particular.

Fearing that such disorder would continue or spread, Congress passed the Espionage Act in June, which was intended to prevent interference in wartime operations and recruitment, prohibit insubordination within the military, and prohibit any form of support for US enemies during war. Perhaps the most famous application of the Espionage Act against a German-American target was the indictment and conviction (though ultimately overturned by the Supreme Court) of sitting socialist Congressman and anti-war activist Victor Berger.

Using the Espionage Act legislation, the federal government enacted an expanding series of regulations for aliens residing in the United States. The first such regulations were implemented immediately at the outbreak of war and stipulated that alien enemies may not be in possession of any item that could be used for espionage or sabotage, including aircraft, firearms, or shortwave radios; they could not publish derogatory comments about the US government or military; they could not live or work in areas designated as "prohibitive" by the federal government; they could be relocated to any location designated by the federal government; they must have permission to leave the country; and they must obtain a registration card and keep it on their person at all times (McElroy, 2002). In order to establish even further degrees of control over suspect populations, in November 1917, additional regulations were added wherein enemy aliens were restricted as to how closely and under what circumstances they could be near certain facilities, such as docks, warehouses, or railroads—a provision which concomitantly restricted their employment; they were banned from entering Washington, DC and also from air travel in general; and the US attorney general was given the authority to regulate the movements of enemy aliens who must report at any time or place as specified by authorities, the latter provision having been the primary catalyst for the internment of German-Americans and others (McElroy, 2002).

Requirements were also imposed related to the finances and personal property of German nationals residing in the United States. The Office of Alien Property Custodian was created under the auspices of the US attorney general. Persons deemed "enemy aliens" were required to fully disclose to the property custodian all of their personal and business finances, including bank accounts, as well as provide an inventory of all property they owned. Using the scant justification that property seizure prevented enemy aliens from supporting powers at war with the United States, the federal government in turn seized much of this property totaling around half a billion dollars—roughly equivalent to the annual pre-war federal budget (Gross, 2014). At war's end, seized properties were auctioned to US citizens rather than returned, a policy that was ultimately upheld by the US Supreme Court as legal under wartime laws.

The Sedition Act of 1918 was essentially an expansion of the Espionage Act, which facilitated further government control over potential dissent engaged in by German-Americans or others, via threat of felony prosecution. The Sedition Act prohibited disloyal or derogatory speech either by individuals or media outlets directed toward the US government, military, or flag, with violators facing fines of up to $10,000 and/or a maximum of 5 years in prison. While President Wilson indicated that he could "imagine no greater disservice to the country than to establish a system of censorship that would deny to the people . . . their indisputable right to criticize their own public officials," he simultaneously believed that his administration should possess the ability to censor information that was in the interest of wartime national security (Wilson Presidential Library, 2017). German nationals residing in the United States who failed to comply with the new requirements, or who were otherwise deemed potential security risks, were subject to the possibility of internment. Although many believed the concept and often arbitrary enforcement of the Sedition Act to be unconstitutional, it survived a US Supreme Court challenge in 1919 only to be repealed by Congress 2 years later as public sentiment finally began to relent in terms of favoring nativism and repressing dissent (Hagedorn, 2007).

Internment of "Enemy Aliens"

Even before the US entry into the war, many nationalist-patriotic organizations, such as the American Defense Society and National Security League, were calling for the internment of non-US citizens of enemy countries and those who sympathized with US enemies (Meyer, 2017). Supporters of such a policy as an appropriate security measure often drew parallels with and cited the precedent of the internment of potentially disloyal populations in Cuba and the Philippines during the Spanish-American War as well as the model of Native American reservations as a vehicle for confinement of elements perceived to be hostile to US interests (Gardner, 2016). With entry into the war, even many prominent leaders not necessarily known previously for xenophobic hysteria now partook in such rhetoric and advocated stringent measures against enemy aliens, including internment. President Wilson publicly proclaimed that he was "sure that the country was honeycombed with German intrigue and infested with German spies," though little evidence existed to support that perception (Meyer, 2017, p. 149).

The demographics involved illuminate why anti-German paranoia did not result in large-scale internment of German-Americans. As of the 1910 census, 8,282,618 persons listed Germany as their primary country of origin, of which around 2.5 million had been born in Germany (Franck, 1989). Many millions more were either partly German in ancestry or had forbearers who arrived in the United States so many generations previously that they may not have retained any cultural connection beyond their surname. In short, although anti-German sentiment increased to near hysteria in parts of the United States and calls for internment were publicly expressed, there were no practical means of doing so with such a large population. O'Connor states that "the task of interning more than eight million citizens would have been insuperable as guarding them would have taken half the army and would have taken their labor from the war industries" (O'Connor, 1968, p. 377). Additionally, the German-American community wielded considerable political influence both regionally and nationally, with numerous members of the US House and Senate being of German ethnicity.

While undertaken on a modest scale without the much larger population of US citizens of German descent being targeted, a formal federal program of oversight, arrest, and internment of many German-born persons was ultimately implemented upon US entry into the war. On the same day war was declared against Germany, President Wilson proclaimed that German nationals living in the country without US citizenship were "alien enemies" and subject to regulation by the federal government (German Historical Institute, 2017b). As of 1917, male German nationals over the age of 14 were legally required to formally register as aliens with their local postmaster and immediately report any change in address or employment, with the requirements subsequently extended to women and expanded to require fingerprints be taken the following year. Over 250,000 people registered, among whom tens of thousands were questioned or otherwise investigated, and some 6,300 arrested, with smaller numbers sent to wartime internment camps (Krammer, 1997).

Ultimately, a total of 2,048 German aliens, and in some cases their spouses who were US citizens, were interned at two camps, usually after having been temporarily confined in other facilities, including local jails. Fort Oglethorpe, Georgia was designated as the internment site for those living east of the Mississippi River and Fort Douglas, Utah for those west of the river (Krammer, 1997). In addition to housing German émigrés designated as enemy aliens by the US government, both facilities also housed German military prisoners of war and merchant sailors who had been aboard German-flagged naval or commercial vessels in US ports or waters at the time war was declared, as well as small numbers of German businessmen or other travelers caught off guard by the entry of the US into the war. Small numbers of other aliens of various European ethnic backgrounds, including Poles, were also detained in the two camps, usually for having been accused of connection to socialist organizations, espionage, or verbal sedition, though as was also the case with German "enemy aliens," actual evidence of such activity was often exiguous and minimal due process was provided.

Although their daily lives were regimented, internees in the camps were in general treated well, having opportunities to plant gardens to supplement their diets with fresh vegetables; read; write letters (subject to censorship) to those outside the camps; attend lectures, language classes, and a range of other educational outlets; and engage in recreational pursuits, including a variety of organized athletic events, musical concerts, theatrical performances, regular film screenings, and the keeping of pets (Depken & Powell, 2009; Ford, 2008). However, both camps were partitioned into different sections for different types of domestic or POW detainees, and the experiences of those interned varied according to classification and status. For example, each camp had separate barracks and rules for wealthy detainees who were allowed to purchase better quality and volume of food and certain other comforts. They were also permitted to retain other prisoners as servants and to abstain from chores at the camps if they hired other prisoners to take their place (Doyle, 2010).

However, conditions were not uniformly idyllic for all internees. Many prisoners of modest financial means were utilized under guard as laborers in various capacities, such as farming or road maintenance in the local area, for which they were paid 25 cents per day in vouchers that could be used at the camp exchange (Bowles, 2007). However, some question exists as to the degree to which this often physically-demanding labor was voluntary. Many internees refused orders by camp officials to sign documentation attesting that they were volunteering of their own free will to work as laborers and were in turn confined in a separate section of the camp and received limited rations as punishment (Gaffney, 1931). The latter situation was perhaps the leading source of prisoner complaints to Swedish and Swiss observers, the Red Cross, and US government officials during the war, but neutral observers, including a Swiss delegation that inspected Fort Oglethorpe, concluded that internees were adequately cared for and fed (O'Bryant, 2009). Although the exact number of fatalities is not known, many internees succumbed to the influenza outbreak which also rampaged through the US civilian and military population at the end of the war.

Among the more controversial aspects of the internment program was the prolonged detention of prisoners at the conclusion of the war. Although the armistice ending the war was signed in November 1918, the majority of civilian "enemy aliens" and POWs were not repatriated to Germany until June and July of the following year, and hundreds of civilian internees, including perhaps the most famous prisoner, Dr. Karl Muck, German national and former conductor of the Boston Symphony Orchestra, remained interned until 1920 (Depken & Powell, 2009). Some internees who had immigrated to the United States and intended to stay were repatriated back to Germany or other countries against their wishes, though in other cases many internees in such circumstances were allowed to remain in the United States.

Institutional Xenophobia

In addition to anti-German popular sentiment, xenophobia also manifested in public policy at all levels of government and in various societal institutions, including academia. Many researchers in the United States and other allied countries refused to cite or even read the scholarship of German-born researchers (Badash, 1979). George Hale, foreign secretary of the National Academy of Sciences in the United States, asserted that scientists of German heritage believed in the innate superiority of the culture of their European ancestors, which contributed to a desire for domination and to their having been "hypnotized into supporting the murderous policies of the German government" (Badash, 1979, p. 107). Professors, including several of German descent (e.g. William Schaper of the University of Minnesota) at a number of US universities, were censured or fired due to their perceived lack of support for the war effort (Wilcox, 1993). In perhaps one of the most extreme examples, the German Department at the University of Michigan was almost disbanded during the war when six of its professors, half the faculty in the department of which most of whom were of German ancestry, were summarily discharged over spurious allegations of disloyalty. The first of the dismissals came in October 1917, with the remaining five firings occurring in March of the following year under the auspice of a "reorganization" of the department, stemming from prominent nationalist alumni and their supporters lobbying the university's board of regents. They alleged the professors used their classrooms for promoting pro-German propaganda and that they had made "unpatriotic or seditious remarks" publicly during wartime (Wilcox, 1993, p. 62).

One of the most significant and longest-lasting institutional impacts of wartime xenophobia was the near eradication of German language from the US educational system. Prior to WWI, German was the most commonly studied foreign language in US secondary schools, with around one fourth of American students studying the language, but by 1922, less than 1% of US high school students enrolled in German language classes (Kamphoener, Helbich & Sommer, 1991). Following US entry into the war, many states and local school districts formally banned the teaching of German, actions which were later ruled unconstitutional by the Supreme Court in 1925. However, due to lingering anti-German sentiment—which would resurge during WWII, German-language classes would never again regain prominence in American schools. Presently, among US students who take a foreign language, only around 4% study German in K–12 schools and 5% at the postsecondary level (ACTFL, 2011; Goldberg, Looney, & Lusin, 2015).

In addition to the actions of state and local school authorities, innumerable actions were taken at the state and local level and by nongovernmental and community organizations against vestiges of German culture in the United States. Iowa, Nebraska, and numerous local governments across the United States passed ordinances restricting the speaking of German in public places or via telephone (Boundless, 2016; DeWitt, 2017). In communities across the United States, the German-derived names of businesses, streets, and other place locations—or even entire communities such as Berlin, Michigan which became "Marne" in honor of the WWI battle—were changed. Many public libraries either of their own accord or at the behest of local governments or civic groups withdrew German-themed books, which were in turn often burned in public ceremonies attended by enthusiastic crowds (Kirschbaum, 2014). Citing fear of sabotage as justification, the American Red Cross banned persons with German surnames from membership. Often resulting from pressure from local government or civic groups, concert halls ceased playing the music of Beethoven, Brahms, Wagner, or other German composers, with the sheet music sometimes being burned in public (Luebke, 1974).

Concluding Observations

Manifestations of nativism and xenophobia in the United States occurred prior to the onset of WWI and targeted various groups at different junctures in US history, such as the vitriol, prejudice, and bigotry often directed toward Native Americans throughout national history and Irish immigrants or Catholics in the 19th century. Nor did WWI mark the final stain made by xenophobia upon the fabric of American society, as it has manifested in some form to some degree in nearly every generation, and we can witness it today in the form of anti-immigrant rhetoric, for example. However, the anti-German sentiment characteristic of the WWI era can be viewed as somewhat distinct in terms of the intense, open, and widespread manner in which it manifested, which has arguably been unsurpassed since. Also, the anti-German backlash marked one of the few occasions in the modern era in which xenophobia became formally and widely codified and institutionalized. The latter processes, whether through codification of laws facilitating persecution without proof of wrongdoing or through less officious machinations on the part of local or civic organizations intended to erase the influence of a large segment of US society, were rationalized at the time via the security concerns of a nation during wartime. Though the scale and the public rhetoric remained somewhat more muted, similar justifications for xenophobic attitudes have since manifested, such as the anti-Japanese-American sentiment and policies of WWII and the anti-Muslim undercurrent which became more prevalent in the wake of 9/11. While security measures in the face of threats to national security will always remain necessary, future security needs should be addressed in a manner that is constitutional in nature and tempered with rationality and a sense of humanity derived in part from an understanding of previous transgressions.

References

American Council on the Teaching of Foreign Languages (2011). "Foreign Language Enrollments in K–12 Public Schools." https://www.actfl.org/sites/default/files/pdfs/ReportSummary2011.pdf Accessed July 16, 2017.

Badash, Lawrence (1979). "British and American Views of the German Menace in World War I." *Notes and Records of the Royal Society in London* 34 (1), 91-121.

Bowles, Edmund (2007). "Karl Muck and His Compatriots: German Conductors in America during WWI and How They Coped." *American Music* 25 (4), 405-440.

Boundless(2016)."TheAnti-German Crusade." *US History Boundless*, November14,2016.https://www.boundless.com/u-s-history/textbooks/boundless-u-s-history-textbook/world-war-i-1914-1919-23/the-war-at-home-180/the-anti-german-crusade-988-8773/ Accessed August 27, 2017.

Davis, Marni (2014). *Jews and Booze: Becoming American During the Age of Prohibition*. New York: NYU Press.

Depken, Gerry and Powell, Julie (2009). *Fort Oglethorpe*. Mt. Pleasant, SC: Arcadia Publishing.

DeWitt, Petra (2017). "The German-American Experience in Missouri during World War I."http://missouriorverthere.org/explore/articles/the-german-american-experience-in-missouri-during-world-war-i/ Accessed August 28, 2017.

Dobbert, G.A. (1967). "German-Americans Between New and Old Fatherland, 1970-1914." *American Quarterly* 19 (4), 663-680.

Doenecke, Justus (2011). *Nothing Less Than War: A New History of America's Entry Into World War I*. Lexington, KY: University of

Kentucky Press.

Doyle, Robert (2010). *The Enemy in Our Hands: America's Treatment of Enemy Prisoners of War from the Revolution to the War on Terror.* Lexington, KY: University of Kentucky Press.

Eardley, Nick (2014). "Files show confusion over Lusitania sinking account." BBC http://www.bbc.com/news/uk-27218532 Accessed July 30, 2017.

Fiebig-von Hase, Ragnhild and Lehmkuhl, Ursula (1997). *Enemy Images in American History.* New York: Berghahn Books.

Ford, Nancy (2008). *Americans All: Foreign-born Soldiers in World War I.* College Station, TX: Texas A&M University Press.

Frank, Irene (1989). *The German-American Heritage.* New York: Facts On File.

Gaffney, Thomas (1931). *Breaking the Silence; England, Ireland, Wilson and the War.* New York: Liveright.

Gardner, Hall (2016). *The Failure to Prevent World War I: The Unexpected Armageddon.* New York: Routledge.

German Historical Institute (a) (2017). *Percentage of Americans of German Origin.* https://www.ghi-dc.org/research/completed-projects/immigrant-entrepreneurship.html Accessed June 15, 2017.

German Historical Society (b) (2017).*Immigrant Entrepreneurship.* https://www.immigrantentrepreneurship.org/image.php?rec=1531 Accessed August 12, 2017.

Goldberg, David, Dennis Looney and Natalia Lusin (2015). "Enrollments in Languages Other Than English in United States Institutions of Higher Education, Fall 2013." *Modern Language Association of America.* https://apps.mla.org/pdf/2013_enrollment_survey.pdf Accessed July 15, 2017.

Grohsgal, Leah (2014). "Chronicling America's Historic German

Newspapers and the Growth of the American Ethnic Press." National Endowment for the Humanities. https://www.neh.gov/divisions/preservation/featured-project/chronicling-americas-historic-german-newspapers-and-the-grow Accessed June 29, 2017.

Gross, Daniel (2014). "The US Confiscated Half a Billion Dollars in Private Property during WWI." Smithsonian: World War I 100 Years Later. http://www.smithsonianmag.com/history/us-confiscated-half-billion-dollars-private-property-during-wwi-180952144/ Accessed August 26, 2017.

Gullace, Nicoletta (2005). "Friends, Aliens, and Enemies: Fictive Communities and the Lusitania Riots of 1915." *Journal of Social History* 39 (2), 345-367.

Hagedorn, Ann (2007). *Savage Peace: Hope and Fear in America*, 1919. New York: Simon and Shuster.

Kamphoener, Walter, Wolfgang Helbich and Ulrike Sommer (1991). *News from the Land of Freedom: German Immigrants Write Home.* Translator: Susan Vogel. Ithaca, NY: Cornell University Press.

Kazal, Russell (2004). *Becoming Old Stock: The Paradox of German-American Identity.* Princeton, NJ: Princeton University Press.

Kirschbaum, Erik (2014). *Burning Beethoven: The Eradication of German Culture in the United States during World War I.* New York: Berlinica.

Kirschbaum, Erik (2015). "Whatever Happened to German America?" *New York Times.* September 23, 2015.

Krammer, Arnold (1997). *Undue Process: The Untold Story of America's German Alien Internees.* Lanham, MD: Rowman & Littlefield.

Krogstad, J.M. and M. Keegan (2015). "From Germany to Mexico: How America's Source of Immigrants has Changed Over a Century." Pew Research Center. http://www.pewresearch.org/fact-tank/2015/10/07/a-shift-from-germany-to-mexico-for-americas-immigrants/ Accessed

June 1, 2017.

Library of Congress (2017). *German-American Immigrant Entrepreneurship.* https://www.ghi-dc.org/research/completed-projects/immigrant-entrepreneurship.html Accessed June 29, 2017.

Luebke, Frederick (1974). *Bonds of Loyalty: FwGerman-American and World War I.* DeKalb, IL: Northern Illinois University Press.

Manning, Mary (2014). "Being German, Being American: In World War I, They Faced Suspicion, Discrimination Here at Home." *Prologue* Summer 2014, 14-22.

Manz, Stephan (2014). *Constructing a German Diaspora: The "Greater German Empire", 1871-1914.* New York: Routledge.

McElroy, Wendy (2002). "World War I and the Suppression of Dissent." Independent Institute. April 1. http://www.independent.org/newsroom/article.asp?id=1207 Accessed August 17, 2017.

Meyer, G.J. (2017). *The World Remade: America in World War I.* New York: Random House.

New York Times (1917). "Puts No Rigid Ban on Austrians Here." December 13, 1917. Accessed August 12, 2017.

O'Bryant, Jeff (2009). *A Brief History of Catoosa County: Up Into the Hills.* Mt Pleasant, SC: Arcadia Publishing.

O'Connor, Richard (1968). *The German-Americans: An Informal History.* Boston, MA: Little, Brown and Company.

Petty, Antje (2013). "Immigrant Languages and Education: Wisconsin's German Schools." In *Wisconsin Talk: Linguistic Diversity in the Badger State.* Purnell, Thomas, Eric Raimy, Joe Salmons (eds.). Madison, WI: University of Wisconsin Press.

Thompson, J. (2014). *Never Call Retreat: Theodore Roosevelt and the Great War.* New York: Springer.

Trommler, Frank (2009). "The Lusitania Effect: America's Mobilization against Germany in World War I." *German Studies Review* 32 (2), 241-266.

Trommler, Frank (1998). "The Historical Invention and Modern Reinvention of Two National Identities." In *Identity and Intolerance: Nationalism, Racism, and Xenophobia in Germany and the United States*. Norbert Finzsch and Dietmar Schirmer (eds.). Cambridge, UK: Cambridge University Press.

US Census Bureau (1999). "Region and Country or Area of Birth of the Foreign-Born Population, With Geographic Detail Shown in Decennial Census Publications of 1930 or Earlier." https://www.census.gov/population/www/documentation/twps0029/tab04.html. Accessed August 8, 2017.

Wilcox, Clifford (1993). "World War I and the Attack on Professors of German at the University of Michigan." *History of Education Quarterly* 33 (1), 59-84.

Wilson Presidential Library (2017). "Wilson and Censorship of the Press."http://www.woodrowwilson.org/library-archives/wilson-elibrary/explore-featured-documents/wilson-and-censorship-of-the-press Accessed August 17, 2017.

3

American and Russian Geopolitical and Geostrategic Interests During World War I

Raluca Viman-Miller
University of North Georgia

Abstract

This essay looks at the Russian and American positions during World War I, which were fashioned under the great powers' economic interests and the need to reestablish a new world order. As the United States was seeking to expand its interests to the detriment of German expansion, it also sought to befriend the Russians and garner their support. In only this way could Germany be contained and defeated. The Russian Revolution and regime change worked in favor of the Wilsonian arguments against autocracy and for self-determination and participation in WWI, but in the end, Bolshevism proved just as strong an enemy as German and Japanese expansion. This analysis concludes that the competing interests of America and Russia prevented a world order in which the two superpowers were cooperating. Wilson's United States entered the war seeking to improve the world of international affairs and found itself in a world of hardened convictions and new power patterns.

Keywords: United States of America, Russia, Germany, WWI, Russian Revolution, Bolshevism, Wilson

American and Russian Geopolitical and Geostrategic Interests During WWI

The beginning and the ending of any war is marked by the desire to redraw the power lines in the international system. This essay discusses the beginning of the 20th century, particularly the First World War and the beginning of the US–Russian competition for world dominance. WWI proved to be the forecaster of the antagonistic relationship between the two world powers that culminated in the end of the Cold War at the conclusion of the 20th century. After over 4 decades of peace, the Great War was the result of global transformations that did not match the political system of the time. Unfortunately, competing interests did not allow for a peaceful transition.

The breakout of WWI in July 1914 between the Allies (Allied or Entente) and the Central Powers was the consequence of unequal and sporadic economic and political developments from the end of the 19th century to the beginning of the 20th. The Triple Entente was formed by the Russian Empire, the French Third Republic, and the United Kingdom of Great Britain and Ireland; the Central Powers were formed by the German Empire and the Austro-Hungarian Empire, later joined by the Ottoman Empire and the Kingdom of Bulgaria. States such as England and France, old industrial forces and large colonial powers, were surpassed by new states, such as the United States and Germany, which experienced tremendous industrial growth. Such was the nature of these developments that the equilibrium of forces in the world and the balance of power were changed. The new world order could only be reestablished with the use of force.

Mainly, the war was triggered by the imperialistic economic and political attitudes of the European powers which were seeking to subordinate and dissolve the independence of small states. At the beginning of the conflict, it was generally believed that it would be a short war, but it lasted 51 months until November 18, 1918. The Central Powers were opposed by the Entente, which was joined later in the war by Japan, Italy, Portugal, Romania, the United States, Greece, and Brazil, with a total of 32 countries being involved.

The Two Great Powers Before the War

World War I signified both the end of a world order and new beginnings. The new balance of power included new social, political, and economic powers, such as that of the United States, while the Russian Empire was on the verge of changing its own international position. Historians have long debated the reasons for the Great War with very little consensus over the main causes. One consensus, however, seems to hold: this war and its peace process changed the shape of the international system and the power arrangements in the world for the next century. The protagonists of the new balance of power in the 20th century, the United States and Russia, were deeply involved in WWI despite the fact that neither one of the two participated in the whole conflagration. The Russian Empire left early while the United States entered late.

Russia Before World War One

On June 28, 2014, in Sarajevo, the Yugoslav nationalist Gavrilo Princip assassinated the heir to the Austro-Hungarian Empire Archduke Franz Ferdinand and his wife, which triggered the July Crisis and led to an ultimatum from the Austro-Hungarian Empire to the Kingdom of

Serbia. Within a month, the Empire declared war on Serbia, and soon the local conflict was transformed into a global conflagration. The Austrian-Hungarian ambitions were to dismantle the challenges they were facing in the Balkans. Before the assassination of the archduke, a letter addressed to its allies was prepared (but ultimately never sent), stating that Romania was no longer a reliable associate because of a Russo-Romanian summit in Constanta (June 14, 1914) and that Russia was moving toward forming an alliance including Romania, Bulgaria, Serbia, Greece, and Montenegro meant to destabilize the Austro-Hungarian Empire with the ultimate goal of breaking it up and moving its borders westward (Albertini, 1953, p. 534). Russia also participated in the Anglo-Russian naval talk in May 1914 (mirroring the Anglo-French naval talks), pressuring the Germans into believing they were encircled by unfriendly forces. One of the major leading powers in Europe between 1870–1914, Russia was initially capable of backing up its diplomatic initiatives with military force, yet by 1914, Russia was facing severe economic downturns and an economic crisis that changed the balance of power out of its favor. Lieven (1983) observes that Russia found in France a dedicated ally that was investing massively in its economic and industrial development, and this alliance was vital for the position of Russia among the European powers. The Franco-Russian Alliance allowed the Russian Empire to survive economically, continue to supply its military forces with enough resources, and, together with France, counterbalance the rapid German development prior to 1914. At the beginning of the century, Germany was one of the most powerful European states.

The Russian Empire of 1914 was playing its last strong cards, but coupled with its economic crisis and the unprecedented rise of enemies' economic power, it was far from the superpower that it had been during the 19th century. Its last hopes were to create strong alliances, prevent Germany and the Austro-Hungarian Empire from gaining more power, and trust that its internal political turmoil would not weaken its international presence. At the beginning of the century, the Russian Empire was fighting for survival.

This instability was inconsistent with most of its pre-20th-century existence. Much of the Russian Empire's expansion took place during the 17th century and culminated in the conquest of the Pacific basin during the 19th century. By 1914, Catchpole (1974) describes the Russian Empire as a conglomerate of over 170 million people with enormous territorial holdings, one sixth of Earth's total land mass stretching from Finland to the west and to Siberia on the east. Its military might was renowned. St. Petersburg could gather millions of soldiers to fulfill its military needs. During peacetime, Russia kept 1.5 million combatants available on contingent—comprising the largest military in Europe—while during war, this number could easily quadruple or increase five times. When Alexander II came to the throne in 1855, the Russian Empire flourished. He was responsible for the emancipation of the serfs in 1861, an event that, despite its liberal value, weakened the land-owning aristocracy and freed up labor that moved to the cities, promoting industrial development and creating a new working class. He was also responsible for reforming the judicial system, abolishing capital punishment, and promoting universities among other advancements. He was one of the most successful reformers of the Russian Empire since Peter the Great. Alexander II sold Alaska to the United States in 1867, as he felt that the colony was hard to control and weakened Russia's international position. He carried on the Russo-Turkish war as the defender of fellow Christians in the Balkans. Despite his disappointment regarding the peace agreement convened upon at the Congress of Berlin, he respected his international agreements. He was also the emperor who modernized the Russian military complex, according to Moss (1958).

It is important to mention that during Alexander II's reign, the nihilist movement was developing in Russia. This movement originated from liberal philosophical ideas but moved on to create a strong political force involved in calls for reform, such as for a state led by common people. The pushback of the imperial powers against nihilism lead to the creation of the more radical faction of anarchism, and anarchists assassinated the tsar in 1881. Alexander II was followed by his less progressive son Alexander III who, at the domestic level, reversed the liberalization instituted by his father and cut Russia's international ties with the more developed Western powers. Byrnes (1970) states that, influenced by his tutor, this tsar was frightened by such ideas as democracy, freedom of speech or the press, constitutions, or parliamentary systems. Alexander III hunted down members of revolutionary movements and carried out a process of Russification across the whole empire. His father's attempts at modernizing Russia were not continued, thus limiting the Russian Empire's ability to grow domestically. Based on agricultural societies almost exclusively until the mid-1800s with its very little industrial development mainly supported by French investments, Russia was trailing far behind its Western neighbors when it came to economic development. Even if minor, the little industrial development created new problems for the domestically unstable empire. The new urban areas were populated by peasants who came to the cities to get jobs and were met by long working days and poor living conditions, triggering social disruptions and political agitation. From a political perspective, Russia was characterized by dissatisfaction and divisions, creating a fertile ground for revolutionary and anarchist movements. The tsar who believed that he was the representative of God on Earth was leading with an iron fist an empire that did not benefit from any of the modern political institutions already present in other European states. Russia did not have a constitution or an elected parliament capable of exercising any power, and about four fifths of the massive Russian population were peasants, who were usually uneducated, religious, superstitious, and poor. These were the social and political realities of one of the greatest empires at the beginning of the 20th century.

Seton-Watson (1967) remarks that during WWI, the Russian Empire was led by Nicholas II who expended the Russian territories even farther. His conquests only went so far, as he lost the war in Japan, exposing his military weaknesses of poor coordination, poor command, and lack of infrastructure. Also, during Nicholas II's reign, Russia experienced an industrial revolution that created even more social unrest and the necessary forces that ultimately triggered the demise of the empire. Two very important political forces that were extremely relevant for the developments of the day were the Bolsheviks led by Vladimir Lenin and the Mensheviks led by Yuli Martov. Both factions claimed socialist doctrine origins. The Mensheviks believed that the best avenue for change was for the empire to slowly die away and to be peacefully followed by a democratic republic where political forces such as socialist parties and liberal parties could cooperate to the benefit of the people. The Bolsheviks, on the other hand, advocated for a forceful takeover, a breakdown of the old regime to be replaced by a small elite of professional revolutionaries controlled by the party (Manning, 1982). The domestic political situation was becoming increasingly less stable, and as a direct result in January 1905, a demonstration meant to deliver a petition to the tsar was met by a loyal Cossacks force at the winter palace, and the massacre that followed, known as Bloody Sunday, outraged an already excited population, turning the petition delivery into a call for the demise of the empire. The tsar, weakened by these revolutionary events, agreed to a set of reforms that, despite their good intentions, were too little and too late, and because of the internal lack of unity in the revolutionary forces, even these were dropped by 1906.

The Russian Empire's foreign policy continued to be as ambitious as ever. The tsar was clearly competing with the rest of the European powers for more influence in Central Eastern Europe, and he sought to take advantage of the clear forthcoming demise of the Ottoman Empire. Also, as the defender of the Christian Orthodox and Slavonic people in the region, the Russian political forces were encouraging and agitating the Austro-Hungarian occupied territories, butting heads with a bicephalous monarchy. Russia was in no way in a friendly position toward the Austro-Hungarians, while the relationship with Germany was in much better shape. The Russian-German friendship was cultivated by German chancellor Otto von Bismarck and the tsar, and it continued to present very little potential for trouble. A great relationship was also anticipated by the fact that the tsar and Kaiser Wilhelm II were cousins and they got along well. At the time, very little was known about the fact that the kaiser had a very poor opinion of Russian domestic and foreign policies and that the kaiser had no interest in continuing to cultivate good relations with Russia (Llewellyn et al., 2014). At the time of the assassination of Archduke Franz Ferdinand of Austria-Hungary, Russia was considered the defender and supporter of the Serbs, and its mobilization of forces in July 1914 as a result of the July Crisis pitted it not only against the Austro-Hungarian Empire but also against its old friend Germany, which jumped at the opportunity to defend its neighbor to the East. The ultimate reasons for the Russian participation in the war were far deeper than defending Serbian nationalistic values and enforcing an existing alliance or client-patron relationship.

The United States Before World War One

On the other side of the ocean, the democracy of the United States of America was following a different developmental pattern. The new republic was going through rapid industrialization, and it was expanding and growing economically at an unprecedented pace. The second half of the 19th century was the fastest growing period of development, which meant a quick rise in production, earnings, and personal prosperity. In turn, these developments made America one of the most attractive places on Earth for immigrants coming from different backgrounds seeking political and religious freedom as well as economic benefits. These migratory waves were determined not only by the promise of economic prosperity and political freedom but also by the fact that small states in Europe were going through hard times at the hand of the great European empires. The Hungarian allies of the Austro-Hungarian Empire were known for their harsh policy of Magyarization, the religious conversion and ethnic cleansing of small peoples of Central Eastern Europe that were part of the empire. The American population grew tremendously during this time. This second part of 19th century development was matched immediately by the military development of the republic, forecasting the beginning of the American imperialist ambitions. This is the time when the United States began investing in naval technology, including the steam engine and advanced armament as main features of its maritime forces. Domestically, the United States was reconstructing a society that was eliminating slavery and laying the foundation of the modern capitalist economy; it was creating its industry, banking, and financial organizations; it was exponentially increasing the miles of installed railroads as technical developments and production methods improved. In short, the United States was going through its "Gilded Age," as Mark Twain called this period and the Progressive Era (Twain et al., 1964). Nichols and Unger (2017) describe the Gilded Age developments as including new inventions, such as the telephone, telegraph, and the use of electric power. It also meant the development of investment capital markets and the beginning of stock markets as we know them today.

This extraordinary period was not without its problems. The political arena was often accused and proven guilty of electing mediocre figures in its highest positions and allowing the newly enriched tycoons to purchase their political favors. New industrial developments were not keeping up with the expected rights of workers or with the necessary wages to ensure decent living conditions for the working class. Lacking appropriate legislation, women and children were exploited as cheap labor. All these social problems triggered the formation of unions, and in 1877, the United States was paralyzed by a 6-week railway workers' strike. Buenker et al. (1977), however, describe the Progressive Era as a time of great social, political, and economic reforms. The Progressives were using grassroots organizations formed by middle-class citizens that were pushing the United States to find new roles for the government in education, to improve the status of women in society and politics, and to fight against deep-rooted corruption in political organizations. The social and political rights of American citizens were brought into consideration and became salient during this progressive time. Despite not finding success until much later, this is the time when women's rights groups, such as the National Women's Suffrage Association lead by Susan B. Anthony, started their activity.

From a foreign policy perspective, the United States had increased its territory with the purchase of Alaska, but it had no other trans-American ambitions. Its declared interests were isolationist, and it conducted policies meant to prove the lack of interest in involving itself in the European affairs of the time. In 1897, the United States even passed on the opportunity to formally and officially annex the territories of Hawaii despite their desire to become an integral part of the country. Herring (2008) underlines the fact that revolutionary forces in Hawaii eventually were successful at joining the United States in 1898, but Hawaii became the 50th state years later in 1959. The United States was interested in keeping its influence on its home continent and making sure that its powers were growing.

Part of the larger current of independence from the Spanish colonial empire, Cuba sought to gain its own freedom. Coletta (1957) describes the end of the century for the United States as the transition from a regional to a global power. The American public grew sympathetic to the cause of the Cuban people, and as soon as USS *Maine* was sunk near the harbor in Havana, the US government took advantage of the popular support for the confrontation and declared war on the Spanish Empire. The Spanish-American War, which lasted barely 4 months between April and July 1898, brought undeniable evidence of the newfound American power. As a result of the war and the Peace Treaty from Paris, the United States extended its influence over Cuba, Puerto Rico, the Philippines, and Guam. The American naval power was now recognized as one of the greatest conquering forces in the area, and it afforded the new US foreign policy ambitions. The new goals were to gain influence over the Panama Canal and the Caribbean, while the Roosevelt Corollary and the Roosevelt Doctrine self-proclaimed the right of the United States to interfere in the region in order to stabilize weak states, further weakening the influence of the Spanish Empire and establishing US hegemonic ambitions. By 1910, when the Mexican Revolution started, the two countries were experiencing the last few days of over a half a century of peace. American business interests were put in jeopardy, and its territory was experiencing an influx of refugees in the hundreds of thousands, basically requiring the American government to try to secure the Mexican political situation (Llewellyn et al., 2014).

American domestic and foreign policy was simpler and easier to ascertain compared to those of the great European empires. Its growing economic influence and developing military power placed it in the position of a global player by the turn of the century. American isolationist foreign policy and the decision to keep things simple and near home by acting as a hegemon only in North America soon came to an end as US economic interests and technological developments afforded the nation more power and the desire to keep rivals out of their spheres of interest.

The Two Great Powers and World War One
Russia

World War I is not the war most referred to in the history of international relations. Historians more often refer to World War II or the Vietnam War when illustrating major changes in the world order. Yet, World War I is the war that placed the United States in a different role, the role of a leader from not only a political but also an economical perspective. In four years and four months, over nine million soldiers and over seven million civilians died in one of the deadliest conflicts that mankind ever experienced (Willmott, 2003).

The July Crisis culminated with a declaration of war from the Austro-Hungarian Empire to the Kingdom of Serbia, immediately involving the Russian Empire as a defender of the Slavic interests in the region. The Russian Empire started mobilization before the official declaration of war on July 28, and despite the German demand that Russia stand down, it continued to prepare itself for war against the bicephalous empire. Ultimately, this was the German desire, as Germany considered Russia unprepared for war, and the only chance for the survival of its allies, the dual monarchy of Austria-Hungary, was the demise of the Serbian people as part of a larger show of force in the region. The Russian take on these actions was to simply follow its own geostrategic and geopolitical interests. Russia decided on general mobilization in order to maintain its position and prestige in the Balkans and opened a front against both Germany and Austria-Hungary when, on the August 1, 1914, Germany declared war against Russia in support of the dual monarchy.

Lieven (1983) describes the political climate of Europe in 1914 as dominated by a weakened Russian position in the region with very little success in adding more territories and controlling new positions. British and German supremacy led on, spurred by economic development. A deteriorating friendship between the dual monarchy and Russia was mainly caused by competing interests. Russia depicted itself as the liberator of the Balkans while the Austro-Hungarian monarchy was the ultimate conqueror of additional territories. Also, Russia was clearly concerned about seeing its interests attacked by the fact that their post-Balkan War actions to maintain European peace were not followed in kind by the Central Powers. These circumstances were already compounding the fact that Germany was ready to challenge Russian supremacy in the region, and their old friendship, regardless of family relations, was disintegrating.

Despite its political and diplomatic ambitions, the Russian political and economic realities were extremely difficult. The famously large and feared Russian military force, amounting to millions of combatants, was severely underequipped and poorly managed, lacking both organization and logistical coordination. Riasonovsky (2004) describes the Russian military participation on the Eastern Front as a conglomerate of desperate yet tenacious Russian soldiers, comprising an armed force that suffered tremendous casualties and was so poorly equipped that soldiers sometimes were encouraged to find their own weapons. The tsar failed to engage the Russian civilian surge of support for the war, missing the opportunity to reinforce its military strength, and while it was pinning down the Austrian and German forces on the battle fields, it failed to engage the very scope of their declared participation in this war, defending the usurped ethnic and religious minorities of Austria-Hungary. Despite an early successful campaign on the Eastern Front, the Russian offensive against the German-Austrian coalition in particular suffered tremendous losses. The Central Powers created a veritable wall against the Russian invasion on the Polish border, but the biggest obstacle to Russian success was their own lack of technological development and their armament deficiencies, their lack of military discipline among high-level officers, and increasing corruption, which caused more problems than their tactical decisions. The tsar relied very little on the expert advice of such political institutions as the Duma and more on that of Rasputin, a controversial mystical and religious figure whose influence at the court was assured by his close friendship with Tsarina Alexandra.

As already mentioned, Tsar Nicholas II was not a visionary, and his understanding of political matters did not come near that of his grandfather. His misguided ambition to seize more influence and land backfired. His ineptitude, bureaucratic failure, enormous losses on the Eastern Front, and lack of popular support accelerated the foreseeable outcome of the Russian monarchy's downfall in February 1917. Curtis (1998) points out that the war strategies employed by the Germans and the Ottoman forces cut off Russian supplies and main imports. By 1915, the economic crisis—coupled with social unrest generated by the lack of resources—the peasants' desires for land reform, and many strikes among the low paid factory workers, as well as the nobility's distrust of the tsar's ability to lead the affairs of the empire beyond the advice of the self-proclaimed holy man Rasputin, ended with the assassination of this priest, an act which nevertheless failed to recover the Russian monarchy's lost reputation.

Rabinovich (2008, p. 1) argues that "the February 1917 revolution... grew out of prewar political and economic instability, technological backwardness, and fundamental social divisions, coupled with gross mismanagement of the war effort, continuing military defeats, domestic economic dislocation, and outrageous scandals surrounding the monarchy." The tsar reacted by ordering the dissolution of the Duma, ordering the troops to shoot at demonstrators. The soldiers famously sided with the demonstrators, thus sealing the fate of the Russian monarchy forever. Nicholas II abdicated on March 2, 1917, and a provisional government declared a Russian Republic.

Pipes (2011) states that the Bolsheviks, aided by German funds meant to destabilize Russia's monarchy along with its war potential, offered protection to the exiled Vladimir Lenin and supported the overthrow of the moderate provisional government and its replacement by the newly formed soviets—organizations formed by the Bolsheviks in the territories and supported by disgruntled peasants and impoverished industrial workers. This movement became known as the October Revolution. In 1917, Lenin dissolved the last traces of a potential democratic regime by dissolving a newly formed Constituent Assembly, withdrew Russia from the war, and entered a peace agreement with the Germans at Brest-Litovsk, where Russia agreed to surrender major territories from its western borders. This March 3, 1918, Treaty of Brest-Litovsk determined the end of the Russian monarchy and the end of Russian participation in World War I, yet when the Germans lost the war, the Russian government sought to revise the treaty in a desperate attempt to minimize its losses.

The United States of America

The US geopolitical and geostrategic interests prior to the Great War were clearly limited to the regional influences it exercised on the North American continent and its neighboring territories. It had a declared interest in staying out of Europe's rotten affairs, and it fought long and hard to push European influence off the American continent. At the beginning of the war in 1914, the United States declared itself neutral. Backed by American public opinion that initially did not favor the war, President Wilson would have liked to see an end to the European conflict mediated by the United States. Though it backed up the Allies, the United States was not officially involved in defending their war interests. American banks lent large sums of money to the British and the French governments mainly directed at the purchase of raw materials, food, and ammunition meant to support the war. Kennedy (2014) underlines that President Wilson tried hard to prod a peace agreement between the British and Germans based on the geopolitical and geostrategic conditions that existed before the war, but he failed because, as in most of the peace initiatives originated before the end of the war, the belligerents thought that the peace conditions were similar to surrender agreements and that they were likely to gain more from ending the war as victors. These peace initiatives failed to present appealing offers to the parties in conflict, according to Kennedy (2014). Finally, President Wilson lost his long battle against American intervention in the war in April 1917 when he asked the US Congress to declare war against Austro-Hungarian aggression. The American intervention in the Great War gained popularity when public antipathy toward German foreign policy intensified over the 1915 sinking of the British ocean liner *Lusitania*, which was also carrying American passengers. The American war supporters, led by former President Theodore Roosevelt, in concert with the British allies, were already pressuring Wilson to declare the German acts of war as anti-American in order to involve the United States in the war. Roberts (2005) calls these pro-war efforts of the Preparedness Movement the "realist" approach to the international politics of the time—the opposite of the "idealist" Wilsonian view. The Preparedness Movement argued that military strength and economic supremacy were more important than such idealist principles as democracy and anti-authoritarianism. They compared the regular American armed forces with those of the rest of the great powers and concluded that the US forces were dangerously smaller. The competition between the liberal movement and the realist movements of the time was also reflected in the national political parties' support for foreign policies. The Preparedness Movement were all Republicans, mostly believing in the realist assumptions and competing with the mostly idealist Democrat views. Some argue that the pro-war ideas were much closer to home and that there are historical arguments that bring into question some of the members of the American government's desire to remain neutral. For instance, McAdoo, President Wilson's son-in-law who was also the secretary of the treasury during Wilson's presidency, had arguably prepared the United States for a decisive takeover in the global financial market. The United States was the largest European war financier, and McAdoo's financial system decisions saved the American economy and its future allies from collapse at the beginning of the war, shifting the global financial balance from Europe to the United States (Silber, 2007). The antiwar supporters stressed the fact that those who were behind the war propaganda were mainly war profiteers who had economic gains derived from American involvement in the war.

Despite this internal divide, the situation in Europe increasingly worsened. The Russians were pulling out of a war meant to keep the Germans from becoming a global power; the French and the British, despite being financially supported by their American allies, found themselves economically choked and facing the unrestricted submarine warfare waged by the Central Powers. The Germans knew very well that this new submarine warfare and the sinking of American boats would likely drag the United States into the war; however, instead of being deterred by this possibility, they attempted to engage a new ally on the North American continent by offering Mexico, in case of success, territories it had lost during the American-Mexican War. The Germans believed they would manage to starve the British economy before the Americans would be ready to join the war.

The famous Zimmermann Telegram, intercepted by British intelligence, and the sinking of seven merchant ships were the last catalysts, forcing the peace-dreaming President Wilson to ask Congress for a "war to end all wars" (Link, 1972), a request Congress voted upon and passed. The United States joined the Allies on April 6, 1917, in the war against Germany, and on

December 7 against the Austro-Hungarian Empire. Yet, Congress never declared war against the Central Powers as an entity.

The American mobilization efforts were rapid and underwent two stages. Initially, it was based on volunteers, but a second phase was instituted in which the government quickly took control of such necessary resources as railroads, food and fuel control, war labor sources, and production of war-related materials. The American contribution to World War I, though late and small compared to that of other Allies, was decisive for the victory of the Entente. The US Navy, which was technologically advanced, as well as the 1.2 million soldiers sent to Europe by September 1918 balanced the war in the favor of the Allies, thereby ending the war by November 1918, earlier than initially anticipated. Americans arrived in Europe at a 10,000-per-day rate, and they were enthusiastically received in France, giving the Germans cause for concern because they had simply underestimated the effectiveness of these new troops on the European front and were ineffective at slowing the American involvement ("1918: A Faithful Ending").

The End of the War and a New World Order

Wilson believed that this American intervention would mean the end of all wars, which in turn would lead to thriving anti-autocratic regimes around the world and a lasting peace that would dominate world affairs from that victorious day forward. Wilson believed his dream of a world without war and a world order of lasting democratic regimes could be fulfilled not only by the American presence in this war and its consequent victorious outcome but also that this outcome would allow him to dictate the conditions of peace and set the stage for what he thought was going to be a new and successful arrangement. This was a conviction that history notes made him keen to ask Congress to approve American entry in the war. He realized that his dreams could not be fulfilled, and the world could not be made safe for democracy without the intervention of the architect of these ideas in its creation. At 100 years since the beginning of the war, the *Telegraph* declared that "when the US joined the First World War in defense of liberty it established itself as leader of the free world and shaped the Russian revolution" (Bishop, 2014). In other words, as Russia was leaving the war theater, a new power was joining the global games.

The American ambitions expressed by Wilson's idealism were articulated in his famous declaration in front of the Senate in January 1917, asking for a "peace without victory" and an international organization, a league of nations with strong American support. He reiterated in this message his call to arms, ultimately including his dreams in his famous Fourteen Points address on January 8, 1918, before the US Congress. In this address, Wilson expressed his desire to end the old type of diplomatic relations that led to past wars; establish open diplomacy; ensure freedom of the sea; remove trade barriers; and establish a league of nations to promote peace, independence of the members, and territorial integrity (The United States Enters the Great War). He also demanded the restoration of Belgium, independence for Poland, the return of Alsace-Loraine to France, self-determination for former Austro-Hungarian and Ottoman territories, and a sympathetic treatment of the Russians, who were involved at that moment in a civil war. In a nutshell, "a breathtaking pronouncement, the Fourteen Points gave new hope to millions of liberals and moderate socialists who were fighting for a new international order based upon peace and justice" (The United States Enters the Great War).

With a few adjustments, these demands were the basis of the beginning of the peace negotiations that ended the war. Wilson fought hard for these points, but he reluctantly ended up agreeing to compromises on some of his important assumptions about the need for peace in the future, such as a large bill of reparation to be charged to the Germans and Allies interfering in the Russian civil war. In this regard, Wilson reluctantly agreed to send a small number of troops to the Northern territories of Russia in order to protect military supplies and keep an eye on the Japanese, actions which would go against the Bolsheviks. The Treaty of Versailles ended up being a mixture of Wilsonian idealism and "old fashioned vengefulness," (Bishop, 2014) and the idea that Wilson would lead the peace process was soon abandoned. Despite the fact that the US Congress did not ratify the Treaty of Versailles, instead signing separate peace agreements, and refused to participate in the League of Nations, it emerged as the most powerful actor at the end of the war. The previously small army expanded to become the most powerful armed force of the century, and the presidential election of 1920 turned into a referendum on the League of Nations and Wilsonian international arrangements that were clearly defeated by the Republican return to the status quo and neutrality. All this political behavior took place in the wake of the general mistrust of foreigners, fear of Bolsheviks and Communists, and a desire for some of the domestic forces to return the United States to its initial position at a regional level. Yet, once victorious in World War I, the United States felt that its role was to enforce the better political and economic options that it was then experiencing. The social, economic, and political outcomes of the war changed the US position, and Wilson in consequence gave the United States the ability to dream larger. The nation did not suffer much internal destruction as a result of the war; hence, it could recover sooner and start making decisions about its future in the international affairs of the world. Yet, the battle between the idealists and the realists continues to this day.

Russia, on the other hand, suffered tremendously. In just a few short years, the Russian Empire—the most autocratic regime in Europe—collapsed, thereby making room for an even more tyrannical movement in the outcome of the Russian Revolution and the civil war: the Bolshevik Communist regime of Russia. After the collapse of the Romanov dynasty and the return of Lenin with the help of the Germans, Russia lost its grip on a potential transition toward a modern democratic state. All power diverted to the soviets, small local organizations of representatives elected by the workers, peasants, and soldiers. Russia was headed fast towards what was soon to become the Union of Soviet Socialist Republics (USSR) from 1922 to 1991, a federal system with a highly centralized economy and one political party that lead with an iron fist from Moscow. Lenin's coup ensured the 1917 success of the radical Communist group, that is, the Bolsheviks. By 1922, Russia joined Transcaucasia, Ukraine, and Byelorussia, forming the core of the Soviet Union as it existed until 1991. Following the death of Lenin in 1924, Joseph Stalin came to power, and his rule was merciless and ruthless. He instated a domestic reign of terror, eliminating any potential opposition to his claimed Marxist–Leninist ideology and initiating moves to occupy more territories in the nearby vicinity of the USSR. This led to the Molotov–Ribbentrop Pact that split Poland and soon helped bring on the second largest conflagration of the 20th century, the Second World War.

Conclusion

This essay offers a brief overview of the positions of the two greatest powers of the 20th century at the beginning of a new world order, a new arrangement that placed the United States of America at the top of the global hierarchy.

In 1917, the world saw new arrangements. The United States had no other options but to join the First World War if it was going to defend its interests and geostrategic gains. President Wilson realized that to be able to shape the future of the new world order, he needed the United States to get involved in the war. The Russian Empire was giving up its fight against German supremacy, the Ottoman Empire continued to show strength and success at defending its geostrategic status quo in the Middle East, and the Germans were winning important battles on the European front.

As a result of the Great War, the world was presented with two opposing models of political organizations: the American model of materialism and personal liberties and the Communist Russian model of collectivism and group interests. This dualism characterized the rest of the century's international relations, and it gave these two competing powers no chance at a cooperative relationship, excepting the brief intermission during the Second World War when once again they had to defeat a common enemy. Both continued to believe all the way to the end of the century that their model was the better option, and this rivalry shaped the politics of the 20th century all the way to the end of the Cold War. Wilson's reference to the war between democracy and absolutism in front of the US Congress proved prophetic.

References

Albertini, Luigi. 1953. *Origins of the War of 1914*. Oxford University Press, London, 3 volumes 1952–1953.

Bishop, Patrick. 2014. *Birth of the American Century.* https://secure.i.telegraph.co.uk/multimedia/archive/02900/world-war-1-part-9_2900224a.pdf.

Buenker, John D., John Chynoweth Burnham, and Robert Morse Crunden. 1986. *Progressivism*. Schenkman Books. Cambridge MA.

Byrnes, Robert F. 1970. "Pobedonostsev: His Life and Thought" in *Political Science Quarterly*, Vol. 85, No. 3.

Catchpole, Brian. 1974. *A Map History of Russia*. Heinemann Educational

Publishers.

Coletta, Paolo E. "Bryan, McKinley, and the Treaty of Paris." Pacific Historical Review 26 May 1957.

Curtis, Glenn E. (ed.) 1998. *Russia: A Country Study*, Department of the Army.

Herring, George C. (2008). *From Colony to Superpower: U.S. Foreign Relations since 1776.* Oxford History of the United States. Oxford University Press.

Kennedy, Ross. 2014. *Peace Initiatives in 1914-1918*-online. International Encyclopedia of the First World War. Ed. By Ute Daniel, Peter Gratrell, Oliver Janz, Heather Jones, Jennifer Keene, Alan Kramer and Bill Nasson. Issued by Freie Universitat Berlin, Berlin 2014-10-08. https://encyclopedia.1914-1918-online.net/pdf/1914-1918-Online-peace_initiatives-2014-10-08.pdf accessed July 20, 2017.

Link, Arthur S. 1972. *Woodrow Wilson and the Progressive Era, 1910–1917.* New York: Harper & Row.

Llewellyn, J. et al, *"Russia before World War I"* at Alpha History, http://alphahistory.com/worldwar1/russia/, 2014, accessed July 5, 2017.

Llewellyn, J. et al, *"The United States before World War I"* at Alpha History, http://alphahistory.com/worldwar1/united-states/, 2014, accessed July 6, 2017.

Lieven, Dominic. 1983. Russia and the Origins of the First World War. New York: St Martin's Press.

Manning, Roberta. 1982. *The Crisis of the Old Order in Russia: Gentry and Government.* Princeton University Press.

Mosse, W. E. 1958. Alexander II and the Modernization of Russia. English University Press. London.

Nichols, McKbight Christopher and Nancy C. Unger (eds). 2017. *Companion to the Gilded Age and Progressive Era.* First Edition. John

Wiley & Sons Inc.

Richard Pipes (2011). The Russian Revolution.

Rabinowitch, Alexander. 2008. *The Bolsheviks in Power: The First Year of Soviet Rule in Petrograd.* Indiana University Press.

Riasanovsky, Nicholas V. and Mark D. Steinberg. 2004. *A History of Russia.* 7th ed. Oxford University Press.

Roberts, Priscilla. 2005. "Paul D. Cravath, the First World War, and the Anglophile Internationalist Tradition." *Australian Journal of Politics and History* 2005 51(2).

Seton-Watson, Hugh. 1967. *The Russian Empire 1801–1917.* Oxford History of Modern Europe.

Silber, William L. 2017. *When Washington Shut Down Wall Street: The Great Financial Crisis of 1914 and the Origins of America's Monetary Supremacy,* Princeton University Press, Princeton, N.J.

The History Place. 1918: A Faithful Ending. Retrieved from: http://www.historyplace.com/worldhistory/firstworldwar/index-1918.html.

The United States Enters the Great War. Retrieved from: https://www.britannica.com/place/United-States/The-United-States-enters-the-Great-War.

Twain, M., Kaplan, J., & Warner, C. D. 1964. *The Gilded Age: A tale of today.* New York: Trident Press.

Willmott, H.P. 2003. *World War I.* New York: Dorling Kindersley.

"American Pride Will Not Stand in the Way of Efficiency":
Public Opinion on the Amalgamation of US Troops into Allied Armies in 1918

Terri Blom Crocker

Abstract

The debate over whether to deploy American troops on the Western Front in 1918 as part of the general Allied armies or as a separate force caused friction between Britain and France and their new ally, the United States. The British and French, desperate for soldiers to replace those lost during four years of war, wanted to integrate the newly recruited Americans directly into their front line armies, whereas President Wilson and General Pershing insisted that their troops should fight as a separate entity. The two leaders persistently rejected their allies' demands for combined forces until Germany's 1918 spring offensive forced them to agree to the temporary amalgamation of American soldiers into the British and French armies. The agreement by Wilson and Pershing that the Allied armies should fight as a coherent force was driven by American public opinion as well as practical considerations. This chapter will analyze the American public reaction to the political and military maneuvering that eventually resulted in a united front against the German army in 1918 and the eventual success of the Allies on the Western Front.

Keywords: Public opinion, American Troop integration, Alliance reaction

"American Pride Will Not Stand in the Way of Efficiency": Public Opinion on the Amalgamation of US Troops into Allied Armies in 1918

The debate over whether to deploy American troops on the Western Front in 1918 as part of the general Allied armies or as a separate force caused much discord between Britain and France and their new partner in the war, the United States. The British and French, desperate for soldiers to replace those lost during four long years of war, wanted to integrate the newly recruited "doughboys" directly into their front line armies, whereas President Woodrow Wilson and General John Pershing, the head of the American forces, insisted that their troops should fight as a separate entity. Wilson's rationale for a distinct American force was driven by his desire to increase his country's influence in any postwar settlement, while Pershing believed that American forces would never be entirely effective if they were integrated into allied armies[1] (Bruce, 2003, p. 146). Their shared belief in the importance of US soldiers fighting as a separate army caused the two leaders to persistently reject their allies' demands for combined forces until Germany's 1918 spring offensive forced Wilson and Pershing to acquiesce

---

1   It's worth noting that Pershing's pride was on the line as well: his professional and personal ambitions could only be satisfied by leading his country's troops in an American-dominated victory on the Western front. As Robert Bruce asserts in *A Fraternity of Arms: America and France in the Great War*, Pershing "had no intention of being a general without an army." Bruce, Robert B., *A Fraternity of Arms: America and France in the Great War* (USA: University Press of Kansas, 2003), pg. 146.

in the temporary amalgamation of American soldiers into the British and French armies.

While the views of both Pershing and Wilson on the subject of merging American troops into the Allied armies already fighting on the Western Front have been analyzed, the attitudes of the American public are less well-examined. The position of the home front on the compelling question of how its soldiers should be deployed to bring about victory in the Great War has not been scrutinized, and no polling appears to exist on civilian views. However, public opinion on the subject can be inferred from media reaction to two catalytic episodes: Pershing's offer, made in the wake of the German 1918 spring offensive, to brigade American soldiers with French and British troops, and a puzzling—and later retracted—summary of operations issued by the British War Office in May 1918, the "Ottowa dispatch."[2] Following these two events, American newspapers discussed the advantages and disadvantages of American troops being integrated into Allied forces and generally concluded that the overreaching goal of beating Germany was more important than the method employed to reach that result. Consequently, while Wilson, Pershing, and Newton Baker, the secretary of war, were all acting under the assumption "that the American people expected that American soldiers would fight under American command as a distinct and coherent body," America's home front was in fact supportive of the idea that their soldiers should fight how, when, and where they were needed rather than according to a strategy set out in advance by their government and military leaders (Zieger, 2000, p. 92).

After remaining neutral for the first 3 years of the conflict, America entered the Great War in April 1917. Later in that year, the Allies,

---

2   This chapter assumes that, absent public polls providing clear views on matters of national interest, the opinions expressed in newspapers, which could influence public opinion but were more likely to reflect it, can be taken to stand in for public viewpoints. After all, a newspaper that consistently ran against the beliefs of its readers would quickly lose those same readers. Under this reasoning, newspapers with large and ecumenical readerships can be utilized to discern the broad outlines of the opinions of their readerships, and, by choosing papers from a variety of regions, broad national trends of thought.

increasingly aware that fighting the war as separate national armies had hampered their response to the Central Powers' more coordinated military tactics, established a Supreme War Council, and on March 28, 1918, the French general Ferdinand Foch was named supreme commander of the Allied forces. In that same month, Germany began their long-expected spring offensive, utilizing the extra soldiers they were able to transfer from the Eastern Front after Russia had withdrawn from the war to provide them with a tactical advantage over their enemies. With the help of those additional troops, Germany hoped to break through the British and French lines in Belgium and France and force the Allies to sue for peace before American forces could arrive in sufficient numbers to reinforce the armies on the Western Front.[3]

Throughout 1917, America recruited and drafted men into its growing army and began to adapt its industry to war-based production, but lack of progress in training soldiers and a failure to effect any sizable increase in weapons output meant that it made little contribution to the war during its first year of participation. Although the Americans made their intentions regarding a separate army clear from the beginning, the British and French, equally obdurate, argued that American recruits should be shipped to France, trained as quickly as possible by the more experienced Allied officers, and sent with all due speed into the already-existing front lines to join seasoned British and French troops. As Charles Cruttwell noted in his 1933 history of the war, "the Allies could bring weighty reasons to support their desire that the Americans should continue indefinitely as auxiliaries of the Franco-British armies, stationed where [the Allied leaders] considered they could be most usefully directed to help their weary comrades" (Cruttwell, 1934, p.556-557). Those rationales included an ability to impart 3 years of hard-

---

3   As Hew Strachan notes in *The First World War*, "Pershing's insistence on independence seemed to have confirmed the Germans' expectations that the United States army would not make an effective contribution until 1919." Strachan, Hew, *The First World War*, (London: Penguin Books, 2003), pg. 310.

earned experience to the untried American troops, as well as access to the fronts and supply lines that the Allies had already established in France.

However, America's leaders, both civilian and military, had other ideas. As Mitchell Yockelson (2008, p.12) notes in *Borrowed Soldiers: Americans under British Command, 1918*, "President Wilson feared that American influence at peace negotiations would be diminished unless his army provided its own expeditionary force." Wilson, uninterested in military matters, was content to leave management of the American army to Pershing as long as his chief objective of retaining influence in the postwar settlement was achieved. Edward Coffman (1968, p. 169), in *The War to End All Wars: The American Military Experience in World War I*, argues that Wilson and Baker would have been willing to let American soldiers serve with the British and French armies, as long as it did not compromise American power at the peace table, but ultimately left such matters to the military commander. As Coffman asserts, "the President and the Secretary would condone amalgamation, but, since they considered it a military matter, they delegated the decision to Pershing."

Pershing's stance on amalgamation was much more unyielding than that of Wilson and Baker, and his reasons for this adamancy went far beyond a simple calculation about America's postwar power. Pershing was determined to hold back American troops from any major involvement on the Western Front until sufficient numbers had arrived in France to form a separate army. Part of this was almost certainly pride: Pershing no doubt pictured himself at the head of a great American army sweeping across northern France, pushing Germany back to its borders, and saving the day for the Allies—a vision that ignored the changes in the nature of war that had taken place during the first 3 years of the conflict. As Gary Sheffield (2001, p. 252) states in *Forgotten Victory: The First World War, Myths and Realities*, "Pershing was as experienced as any American soldier in 1917, but his background was in small wars and what today would be referred to as counter-insurgency." Pershing's previous military experience left him ill-prepared to understand the realities of an artillery-dominated conflict that forced both sides to practice largely defensive tactics.

As a result of his inexperience, Pershing's views on how to win the war were not entirely realistic. His insistence that "French and British tactics were not aggressive enough" (Bruce, 2003, p.121), combined with his belief that the conflict could only be won by the sort of open warfare that the Allied powers had found so ineffectual for the previous 3 years, provided Pershing with sufficient excuses to reject the idea of integrating his forces into the more experienced British and French armies.[4] In a January 6, 1918, memo to General Pétain, Pershing asserted that "there are many real obstacles in the way" of amalgamating American units into the French lines (Center for Military History, 1989, p. 262). "The differences of language, military methods, and national characteristics above referred to," he went on, "would seriously hinder complete cooperation necessary in combat"(Center for Military History, 1989, p. 262). Bruce attributes Pershing's attitude to the fact that he "apparently believed the stalemate on the western front represented a character flaw in Europeans instead of a new set of conditions imposed by the deadly effects of modern weaponry on the battlefield" (Bruce, 2003, p. 124).

Pershing also insisted that the home front would never tolerate American soldiers serving in foreign armies, even if those armies were its allies. In his January 6, 1918, memo to Pétain, Pershing stated this conviction bluntly: "The American people would not approve of giving up the integrity of our organization and scattering it among French and British units" (Center for

---

4   As Sheffield points out, Pershing's faith in the American preference for open warfare cost the lives of many US soldiers: "The parallels between the American offensives of 1918 and the BEF on the Somme are unmistakable. In both cases enthusiastic citizen soldiers launched clumsy, frontal assaults; in both cases the inexperience of commanders and staffs was all too evident" (Sheffield 2007 p. 253).

Military History, 1989, p. 262). He maintained this belief long after the war, writing in his 1931 autobiography that "the nationals of no country would willingly serve under a foreign flag in preference to their own. The national sentiment was such that we could not possibly enter into such an agreement, except in an extreme crisis" (Pershing, 1931, p. 34-35). Pershing's biographer, Perry, affirmed the general's principle, noting that, according to Pershing, "a sense of national identity, not to mention a presidential directive, demanded a separate American force" (2011, p.146).

However, at the same time that Pershing was insisting that Americans would never tolerate the idea of integrated allied armies, press coverage of the war was already belying his argument. As demonstrated by the media in the early part of 1918, the sentiments of the American public were in favor of having their soldiers involved in the fighting under any guise, including that of combined Allied forces. On January 24, 1918, for example, the *Washington Post* published an editorial that endorsed the idea of a "single front" for the Allies, arguing that the Central Powers were able to dominate the war for the first 3 years of conflict because they fought a highly coordinated campaign, whereas each Allied army had fought as a separate force, resulting in delays in reacting to the strategy of Germany and its allies. The *Post* similarly applauded the creation of the Supreme War Council, which was designed to provide a coordinated effort on the part of the Allies even though Americans would also be subordinate to its powers and directions.

The *New York Times* took a comparable stance on the subject in a February 20, 1918, article which stated approvingly that there was "no doubt" that Wilson was in agreement with "having the Supreme Council determine the mutual dispositions of troops and their employment in unison against the common enemy" ("Wilson for," 1918, p.3). While the *Chicago Tribune*'s editorial focus in early 1918 was on the country's lack of preparedness for war, it is clear that it regretted this tardiness because America had to beg help from its allies instead of being able to offer it. As the paper opined on January 12, 1918, that Americans were not only sent abroad in the first place without adequate equipment but were still "using machine guns and artillery drawn from the supply of our allies, who at that same time are exerting 'every power and every resource' to maintain and increase the supply for themselves" was an obvious source of shame ("Our Present," 1918). Although different newspapers approached the subject from varying angles, the overall message was consistent: in the service of defeating Germany, American forces should be pursuing a unified war effort with the other allies—and doing its share of the fighting.

Additionally, even before Pershing's decision that American soldiers would be amalgamated as needed into the British and French armies, the home front was well aware that their soldiers already at the front in France were serving side-by-side with other Allied troops. On February 18, 1918, the *Washington Post* emphasized American cooperation and coordination with Allied forces. An article, entitled "Thousands of U.S. Troops Now With French in Trenches On a Famous Battlefield," made it clear that American soldiers were fighting along with French troops, while an editorial, "The Allies United at Last," noted favorably how "bit by bit the supreme war council has been invested with real executive and directing powers over all the allied armies" ("The Allies," 1918). The *New York Times*, covering a small battle on the Western Front on March 1, 1918, did nothing to disguise the fact that French and American soldiers were fighting together. "The well-placed American machine guns sent streams of bullets into the advancing enemy, and as the German barrage fire lifted the American artillery quickly laid down a curtain of fire, the Germans retiring without a single prisoner," ("Americans Repel," 1918) the paper reported. "There were no American casualties. Five French soldiers were wounded in the fighting" ("Americans

Repel," 1918. As can be seen by this press coverage, Pershing's dream of a separate army was never a reality on the Western Front, and Americans were well aware of this.

Whether Pershing's insistence on an American army fighting as a distinct force was a result of his egoism, his inexperience with the new type of warfare, his disdain for French and British military tactics, or an "eagerness to champion American initiative and individualism," his obstinacy came up against a number of hard facts in 1918 (Keene, 2010, p. 515). The American army, which had not fought in a war since the Spanish-American conflict in 1898, was a small force when America entered the war in 1917. Vast numbers of men were quickly drafted, but there were few facilities to house them in the United States and few experienced officers to train them. Furthermore, American industry, in spite of government intervention, took far too long to reorganize itself on a war footing, leaving the inexperienced soldiers desperately short of the necessary equipment, particularly long-range artillery, which the wide-reaching conflict demanded[5] (Center for Military History, 1989, p. 116). Even had adequate officers, training, and munitions been available, America possessed few ships that were suitable for transporting them to France, forcing America to rely on British shipping to get their soldiers overseas. Due to the German policy of unrestricted submarine warfare and a blockade of the British Isles as part of their attempt to force Britain out of the war, the British were short of food and needed all their available transport to keep their civilians fed and their army in

---

5   A March 21, 1918 letter from the American II Corps memorializing an agreement with the British over supplies demonstrates how much America relied on its allies, even a year after they had entered the war, to provide it with necessary provisions and equipment. According to the letter, Americans were receiving rolling kitchens, guns, automatic rifles and ammunition, bombs, grenades, rockets and flares, replacements of clothing, mounts and mounted equipment from their British partners. The letter notes, however, that the British rum ration was being omitted from the provisions supplied to the Americans, which doubtless more than a few doughboys regretted. The French similarly agreed to provide guns, artillery, ammunition and training to American soldiers.

France supplied. As a result, the British could see little gain and much loss to themselves in making shipping available to the Americans so that their soldiers could park themselves behind the lines until sufficient numbers had arrived to form a separate army.

Additionally, Pershing had in mind a very different schedule for the war than that of his colleagues leading the British and French armies. He thought of 1918 as the year that America's allies would engage in a holding action, keeping the Germans stalemated on the Western Front, and 1919 as the year the Americans, having massed their troops in Europe, would initiate the grand offensive that would ensure a final and decisive victory—with the Americans as the determining factor in Germany's defeat. As Zieger (2000) puts it, "Pershing and the Allied leaders were working on different timetables. US political and military leaders were convinced that there could be no decisive offensive action on the Western Front until at least 1919" (p. 95). This attitude was reflected in the American press: the *San Francisco Chronicle* published an article on January 13, 1918, stating that the arrival of a million American troops in France would enable the Allies to reassume an offensive war in 1919 ("Allies in 1919," 1918).

Events on the Western Front, however, quickly overtook Pershing's original plan. When the German 1918 spring offensive started, German troops advanced rapidly against both British and French forces. As part of the Allied determination to present a united front to the Central Powers, Foch was declared supreme commander of the Allied Forces on March 26. Two days later, Pershing recanted his earlier opposition and agreed to amalgamation of his troops in the armies of the Allies. Pershing's address to General Foch contained the following stirring declaration:

> I have come to tell you that the American people will hold it a high honor that their troops should take part in the present battle. I ask you

to permit this in my name and theirs. At the present moment there is only one thing to do, to fight. Infantry, artillery, aeroplanes – all that I have I put at your disposal – do what you like with them. (Horne, 1930, p. 101)

What caused Pershing's change of heart? It can only have been the realization that, without American troops to swell the ranks of the depleted French and British armies, there would be no opportunity for a grand sweeping charge in 1919, as waiting that long would mean there was no war left to win. The initially successful German 1918 spring offensive, which relied not on principles of open warfare but rather on focusing on weak points in the Allied lines, appeared dangerously close to breaking through the Western Front stalemate of 3 years and forcing an early end to the war. Under these circumstances, a choice between waiting till the Americans could field their own army and doing everything possible to assist the British and French in their fight against the Germans offered no choice at all—any delay would mean the war could end quickly, with the Central Powers triumphant and the Americans officially part of the losing side.

The American press, in accordance with its previous endorsement of coordinated Allied fighting, immediately applauded Pershing's declaration. The *New York Times* noted that Baker was "delighted" ("Baker backs," 1918) with Pershing's offer, and the *Washington Post*'s editorial board wrote approving that "all the productive powers of this nation are mortgaged to the supreme duty of cooperating with France and England in beating Germany on the western front" ("Aspects of the Struggle, 1918). The *Chicago Tribune* focused on the reaction of American troops, excitedly reporting that "the news that Gen. Pershing had placed them at the disposal of the allies for any duties that the French and British staffs might direct has been received with cheering in our front line trenches . . . The effect of the news has been electrical" ("Yanks go," 1918). The *San Francisco Chronicle*, in an editorial written 10 days later, spoke about the country's "gratification at the thought that America is actually over there and doing things on a bigger scale than the pessimists and pacifists would have had us believe possible" ("Getting them Across," 1918).

If the editors of these newspapers disapproved of Pershing's decision or felt that it hampered the American ideal of fighting as a separate army, no sign of it appeared in their papers' columns. In fact, the *New York Times* continued the favorable coverage of Pershing's decision, noting in an article entitled "Pershing's Troops Ready and Willing" that "American troops will soon be fighting in force, side by side with the French and English in the great battle to push back the German drive"[6] ("Pershing's Troops," 1918). On May 4, 1918, the *Los Angeles Times* expressed the same positive attitude toward the amalgamation of American soldiers into the British and French armies, reporting that "French, British, American, Belgian, Italian or Portuguese—no matter what its natural origin—every division is a part of the army of the Allies, and it will go and fight wherever Foch directs" ("Unity of Command," 1918). The *Boston Globe* echoed these sentiments in an article entitled "Allied Armies Now Absolutely Unified/No Distinction under Foch as to Nationalities" ("Allied Armies," 1918).

As demonstrated by the articles and editorials published in these papers, American civilians had no issue with the concept of its soldiers fighting under Allied rather than American control and approved of the amalgamation of American troops into the Allied armies. It can be argued, however, that this was because they had been presented with no other scenario—that a separate American force, however desired by the country's political and military leaders, had not been proposed as an alternative to amalgamation, and that

---

6   In fact, the Times was well aware that American troops had already been fighting side by side with the French, but the addition of the qualifier "in force" made this news.

as a result of manipulation by the press, Americans had been maneuvered into believing that letting their troops fight on the British and French fronts, under Allied rather than American command, was the only option.

This argument was strongly contradicted some weeks later by the reaction to the "Ottawa dispatch," a British communique which dropped like a large stone into very calm waters, spreading ripples of puzzlement throughout the American press. The report, purporting to come from the "war committee of the British cabinet" and sent through official channels to Ottawa, Canada on May 11th, 1918, was not marked confidential and, therefore, was immediately released to the press. The essence of the dispatch was that the Allied powers, completely confident of their "ability to withstand any drive the Germans can launch" had "decided not to use the American army until it becomes a complete and powerful force" ("Won't Use," 1918). In fact, the *Boston Globe*, quoting directly from the notice, reported that as the German army was clearly "'draining their country dry to force a decision before it is too late,'" the Allies, "'having been given the choice of a small immediate American army for defense or waiting till they are reinforced by a complete, powerful, self-supporting American army . . . have chosen the latter'" ("Won't Use," 1918).

Such a declaration could only mystify the American public. Having been told that the immediate use of American troops to support the French and British armies was the only way to prevent the Germans from quickly winning the war, it was now being informed that the services of American soldiers would no longer be needed and that the Allies could manage very well without them.[7] Other newspapers produced similar articles, varying little in the amount of detail provided, with the *New York Times*' May 13, 1918, headline providing an excellent overall synopsis of the press coverage:

---

7   To put the emergent nature of the German 1918 spring offensive in perspective, at this point in time the Central Powers had pushed the Western front back further in six weeks than the line had moved during the three years of previous fighting, with the German army now threatening Paris.

"Allies Won't Use American Army Until It Is Complete/Meantime Await in Confidence Foe's Imminent Attack/German Guns are Pounding the Lines at Many Points/Allies Had to Choose And Decide to Use Our Army When Complete and Self-Supporting"[8] ("Allies Won't," 1918).

Scrambling to decipher a communication so odd that its inexplicableness was referenced in headlines ("Baker Away, Washington Puzzled," 1918), the American press rushed to make sense of the development and find a silver lining in the abrupt change to Allied policy. The *New York Times* reported from Washington that everyone in the city was delighted with the news, not only because it showed the confidence of the Allies in the face of the German advance, but also on account of "the keen desire of American military men to take the field against the enemy as a distinctly American force" ("Baker Away, Washington Puzzled," 1918). The *Globe*'s report, entitled "Washington Welcomes Decision on U.S. Army," put an even more positive spin on the dispatch's contents:

> From a psychological standpoint, the news is doubly welcome since it will mean much to a morale of the American Expeditionary Force to be [a] homogenous unit, instead of being broken up into several groups. The need for these sacrifices, which President Wilson ordered in causing American troops to be brigaded with those of the Allies, is past. ("Washington Welcomes," 1918)

The *Chicago Tribune* also noted that American military leadership was delighted with the idea of fighting as a separate army although it also reported that "officers of the war department were puzzled by the Ottawa dispatch quoting the British war summary" ("U.S. Officers Elated," 1918). On the same

---

8   The *Globe, Post, Chronicle, Tribune* and *New York Times* all featured similar coverage of the dispatch on May 13, 1918; the *Los Angeles Times* did not mention it until the next day, when it had already been withdrawn.

day, it featured the main point in a front-page banner headline, "Save Yanks To Finish Hun" ("Save Yanks," 1918). These newspapers, which described the process of amalgamation with the French and British armies as a "sacrifice" and emphasized the joy that army leadership took in the change of plans, would seem to prove Pershing's point that the American public was impatient for its army to fight as a separate force. It appears at first glance that national pride was being reasserted by the civilian population on the subject and that the country had previously merely tolerated (for the sake of progress in the war) the idea of its troops being integrated into other armies and fighting under foreign flags.

The reception of the immediate repudiation of the Ottawa dispatch, however, demonstrates otherwise. The *Washington Post* printed the British retraction of the communication as its main headline, stating that the American army would, as previously arranged, stay in the front lines, and proclaiming that the dispatch had been issued in error ("Pershing's Men," 1918). The *New York Times* noted that while the statement had sounded authentic, everything contained in it "was contrary to the formal arrangement made by the British, French, and American Governments" for brigading American troops with the French and British armies ("Denies a Change," 1918).

The *Los Angeles Times* focused on the fuss stirred up by the report's contradiction of stated policy, noting that it had "excited much comment at Washington, for it was understood that the American forces had been placed at the disposal of Gen. Foch, commander-in-chief of the Allied forces in France, by Gen. Pershing, the American commander, and it was known that a large number of men now are holding positions in the Montdidier sector of the Somme front" ("Americans Fight," 1918). Aside from the general air of puzzlement caused by the unexpected change in the Allied agenda and the oddity of the Ottawa dispatch, none of these newspapers displayed any disappointment the next day at the reinstatement of the policy of brigading American soldiers with British and French troops.[9] The only sign of any displeasure came in a *New York Times* article that noted that "officers of the United States Army would be gratified if the situation permitted our forces to fight as a separate unit, but all their information indicated that no change of plan was contemplated" ("Denies a Change," 1918).

Continued coverage of the controversy was evenly divided between attempts to find an explanation for the dispatch that had so bizarrely contradicted current Allied strategy, and approval that the policy of brigading American troops with the British and French armies would continue. A *New York Times* editorial dated May 14, 1918, entitled "The Ottawa Mystery," noted that the reasons given for not using the American army until it could take the field as a separate force, in light of the recent great losses on the Western Front, "were a strain on credulity" ("The Ottawa Mystery," 1918). In fact, the paper asserted, a "more illogical and specious document could hardly have been conceived of. It flattered a natural sentiment for independent and decisive action by an American army under an American commander, but at the expense of common sense" ("The Ottawa Mystery," 1918). The

---

9   While the British, in issuing a denial of the contents of the dispatch, provided some half-hearted and obviously untrue statements about the communication—such as one claiming that "owing to the error in compilation it was not made clear that the reference related to the period when America first joined the war and had no relation to present events whereby the American Army is being brigaded with the Entente Armies," which was clearly nonsense, as the Ottawa dispatch referenced Foch as the supreme Allied commander of the Allied armies, which he had not been when America joined the war—no reasonable explanation for the dispatch was ever given, and no historiography referencing the subject has been located ("Story on American Army Incorrect/Statement was Given Out Today at Ottawa/Blunder was 'Due to an Error in Compilation,'" *Boston Globe*, May 15, 1918). Some American papers speculated that the "mix-up" related to Lloyd-George's recent political misfortunes (only a few days before the dispatch was released, he had faced a no confidence vote in Parliament), but it is difficult to see how the contents of the dispatch could bear any connection to the issues he was facing. The Canadian Prime Minister provided a statement to their Parliament a few days after the dispatch, noting only that they had released what had come over the wire from Britain, and stating, in essence, that wherever the screw-up had initiated, it was not Canada's fault. For the moment, the reason for the Ottawa dispatch, so out of keeping with Allied policy, must remain a mystery.

New York Times then pointed out that the practice of brigading American troops with their more experienced allies was not only desirable in light of the current emergency, but "good, practical tactics," as the amalgamation of forces would be the best method "of bringing out the sterling qualities of the American soldiers and accustoming them to the stern ordeal of European warfare" ("The Ottawa Mystery," 1918).

The *Washington Post*'s editorial department was similarly supportive of the "return" to the original policy of brigading American troops with the British and French armies. Their May 16, 1918, editorial, "Ready for any Duty," noted that while the "origin of" the bizarre communication was "a mystery that remains unsolved," it was clear that the original policy of placing untested US soldiers with more experienced Allied armies "is to be followed out" ("Ready for," 1918). The editorial described the Ottawa dispatch as a "blunder" and expressed approval of the original plan of amalgamating American troops with the French and British forces ("Ready for," 1918). It went on to note that it would also be fine if Foch later decided to keep the Americans in reserve for later use (if it fit in with the Allied plans), although waiting to get into combat would frustrate many soldiers, as well as those on the home front. "The Americans going to France are anxious and eager to get into the fray," the *Post*'s editorial affirmed, "and it might be a keen disappointment to American pride if they were obliged to stand back during the big battle and content themselves with preparatory drilling. But American pride will not stand in the way of efficiency" ("Ready for," 1918).

Both the *San Francisco Chronicle* and the *Chicago Tribune* featured editorials that stressed the sensible nature of the policy of brigading American troops with their allied counterparts. The *Tribune* noted that although some American military leaders might prefer a separate army, it was "confident [their wishes do] not express the spirit of Gen. Pershing or Gen. March. 'National' pride' and the 'ambitions of American officers of high rank' are not legitimate factors in any decision as to the employment of our troops abroad" ("American Troops," 1918). The *Chronicle* further argued that it was silly to say that American soldiers couldn't reach their "highest efficiency" under French or British army systems, under which other forces, such as the Canadians, had been fighting for years. Any critique that Americans would lose their "'dash and initiative'" by training for trench warfare was also dismissed as nonsense. "What would they have the American Army do?" the *Chronicle* asked rhetorically. "Is it that they want American soldiers to engage in a premature counter offensive of their own, going over the top and beginning an independent drive for Berlin?" ("Trench Warfare," 1918). The paper followed this editorial a few days later with a full-page drawing showing the Allied armies united behind Belgium, characterized as "the principle of freedom of small nations to rule themselves," and represented as a child leading the consolidated forces ("And a little," 1918).

Wilson and Pershing may have had firmly held tactical and personal reasons for their desire to have a separate American army, but their assessment that American national pride was at stake was far off the mark. The home front was aware from the start that American soldiers were fighting in conjunction with British and French forces rather than as a distinct army. Pershing's announcement in March 1918 that American troops would be brigaded with other allied soldiers elicited only cheers that the United States would finally become active in the common goal of defeating the German army. And finally, when the news that Americans were to sit out the 1918 spring offensive and let the Allies take care of holding back the German advance was put before the American public in the form of the hastily retracted Ottawa dispatch, the response was a collective national shrug and, in some cases, indignation that the American soldiers would be missing a good fight. When the error was corrected and the announcement made that American soldiers would indeed be brigaded with the French and British armies as per Pershing's March 1918 offer and, furthermore, fighting soon, the public's reaction was to applaud the policy's good sense and reiterate the sound reasons for such a course of action. While there is little doubt that military leadership, as quoted by these newspapers, would have liked American forces to remain separate, such a preference was not shared by the home front, which had indicated approval of the process of amalgamating American soldiers into French and British forces long before the policy was proposed, and were quite happy to continue to support such a course of action even after the prospect of a distinct American force was dangled in front of them.

The attitudes of the American home front, as expressed through these papers, to the news that its soldiers were indeed going to be sent to fight with British and French troops can best be summed up by the final line of the *Washington Post*'s editorial on the subject of the Ottawa dispatch: "Whatever is decided to be best for the allied cause the Yankee army will do willingly" ("Ready for," 1918). The policy of brigading American troops with Allied armies, and having them fight under foreign flags, reluctantly agreed to by Pershing and reported as a "disappointment" to US military leadership, was viewed very differently by the American public, which appears to have been less motivated by notions of national pride than its leaders believed. Pershing's ambition to lead a separate American army, as well as Wilson's desire to preserve America's influence at the postwar bargaining table, have been well-documented. What is less often noted, however, is the attitude of those on the American home front, who, while they certainly wanted the Allies to win the war, cared much less than Wilson and Pershing whether Americans would get the credit for the victory.

References

Allied Armies Now Absolutely Unified/No Distinction Under Foch as to Nationalities/Every Recent Enemy Effort to Feel Out Their Power Has Met a Setback. (1918, May 12). *Boston Globe*.

Allies in 1919 Will Be in Position Again to Assume Offensive/With Coming of Million American Troops, Germans Will Be Outmanned, and Advantage Will Pass for Entire Period of War to Their Enemies. (1918, January 13). *San Francisco Chronicle*.

Allies Won't Use American Army Until It Is Complete/Meantime Await in Confidence Foe's Imminent Attack/German Guns are Pounding the Lines at Many Points/Allies Had to Choose And Decide to Use Our Army When Complete and Self-Supporting. (1918, May 13). *New York Times*.

Americans Fight Now/Not to Wait Until Army Powerful/Statement Sent Out from Ottawa is Declared in London an Error/Secretary Baker Issues a Denial and with it an Explanation, Also. (1918, May 14). *Los Angeles Times*.

American Troops Abroad. (1918, May 1). *Chicago Tribune*.

And a Little Child Shall Lead Them' To Victory. (1918, May 19). *San Francisco Chronicle*.

Aspects of the Struggle. (1918, April 2). *Washington Post*.

Baker Away, Washington Puzzled. (1918, May 13). *New York Times*.

Baker Backs Offer of Army/Expresses His 'Delight' at Pershing's Request to Co-operate/Enthusiasm in the Ranks/ Secretary Says Men, Glum Over Inaction, Threw Hats into Air and Danced. (1918, March 31) *New York Times*.

Baldwin, H. W. (1962). *World War I: An Outline History*. Grove Press.

Brophy, J. (1936) *The Five Years: A Conspectus of the Great War Designed*

Primarily for Study by the Successors of Those Who Took Part in it and Secondarily to Refresh the Memory of the Participants Themselves. Arthur Barker Ltd.

Bruce, R. B. (2003). *A Fraternity of Arms: America and France in the Great War*. University Press of Kansas.

Buchan, J. (1919). *Nelson's History of the War, Vol. XXII: The Darkest Hour*. Thomas Nelson and Sons.

Carey, G.V. and Scott, H.S. (2011) *An Outline History of the Great War*. Cambridge University Press.

Coffman, E. (1968). *The War to End All Wars: The American Military Experience in World War I*. The University Press of Kentucky.

de Chambrun, J.A. and de Marenches, C. (1919). *The American Army in the European Conflict*. New York: MacMillan Company.Cruttwell, C.R.M.F. (1934). *The Great War 1914-1918*. Clarendon Press.Falls, C. (1959). *The Great War*. G.P. Putnam's Sons. (1930).

Denies a Change in Plan for Army/Our Troops Will Continue to be Brigaded with Allies, Says Baker/Story Amazes Reading/ Envoy Sure That Ottawa Report Did Not Come from British War Cabinet. (1918, May 14). *New York Times*.

Getting Them Across. (1918, April 10). *San Francisco Chronicle*.

Keegan, J. (2001). *An Illustrated History of the First World War*. Random House.

Keegan, J. (1999). *The First World War*. Hutchinson, 1998, reprinted New York: Alfred A. Knopf, Inc.

Keene, J.D. (2010). "The United States". *A Companion to World War I*. (J. Horne, Ed.). Blackwell Publishing Limited.

Mead, G. (2002). *The Doughboys: America and the First World War*. Overlook Press, 2002.

Our Present and Immediate Task. (1918, January 12). *Chicago Tribune*.Save

Yanks to Finish Hun. (1918, May 13). *Chicago Tribune*.

Perry, J. (2011). *Pershing: Commander of the Great War*. Thomas Nelson, Inc.

Pershing, J.R. (1931). *My Experiences in the World War, Vol. II*. Frederick A. Stokes.

Pershing's Men Stay on the Battle Front/Reports of Withdrawal Plan Error, London Says/'Official,' Says Borden/Baker and Reading in Statements Say Opposite is True/'Committee a Mystery. (1918, May 14). *Washington Post*.

Pershing's Troops Ready and Willing. (1918, April 1) *New York Times*.

Pollard, A.F. *A Short History of the Great War* (reprint, no publisher or publication date listed).

Ready for any Duty. (1918. May 16). *Washington Post*.

Richard, T. (1963). *The First World War* (M. Kieffer, Trans.). G.P. Putnam's Sons.

Sheffield, G. (2001). *Forgotten Victory, The First World War: Myths and Realities*. Headline Book Publishing.

Sheffield, G. (Ed.). (2007) *War on the Western Front*. Osprey Publishing.

Stokesbury, J. L. (1981). *A Short History of World War I*. Harper Collins

Strachan, H. (2003). *The First World War*. Penguin Books.

The Allies United at Last. (1918, February 18). *Washington Post*.

The Ottawa Mystery. (1918, May 14). *New York Times*.

The 'Single Front' Theory. (1918, January 24). *Washington Post*.

Thousands of U.S. Troops Now With French in Trenches On a Famous Battlefield. (1918, February 18). *Washington Post*.

Trench Warfare/ Washington Strategists Offer Very Pronounced Opinions Upon Its Value. (1918, May 15). *San Francisco Chronicle*.

United States Army. (1989). *United States Army in the World War, 1917-1919: Training and Use of American Units with the British and French,*

Vol. 3. Center of Military History.

Unity of Command Helps Allies Win/Ten Million Men Fighting Under Foch; National Lines Wiped Out. (1918, May 4). *Los Angeles Times.*

U.S. Officers Elated. (1918, May 13). *Chicago Tribune.*

Washington Welcomes Decision on U.S. Army. (1918, May 13) *Boston Globe.*

Wilson for Unified Power/No Secret in Washington of His Attitude as to Versailles. (1918, February 20). *New York Times.*

Winter, J. and Bagget, B. (1996). *The Great War and the Shaping of the Twentieth Century.* Penguin Books.

Won't Use Pershing's Army Until Fully Organized; French Gain Near Kemmel; Italians Victors; Active on U.S. Front/Allies Confident of Ability to Stop Huns Alone/Officials Here Pleased/Great Sacrifice to Brigade U.S. Forces with Others. (1918, May 13). *Washington Post.*

Yanks Go To Battle/100,000 Men To Join Foch On Somme Line. (1918, April 1). *Chicago Tribune.*

Yockelson, M.A. (2016). *Forty-Seven Days: How Pershing's Warriors Came of Age to Defeat the German Army in World War I.* New American Library.

Yockelson, M. A. (2008). *Borrowed Soldiers: Americans under British Command, 1918.* University of Oklahoma Press.

Zieger, R. (2000). *America's Great War: World War I and the American Experience.* Rowman & Littlefield Publishers, Inc.

5

Lying, Spying, and Right Defying: The Espionage Act of 1917 and the US Wartime Qualification to the Freedom of Speech

Ashlee Beazley[1]
KU Leuven

Abstract

When it was first passed on June 15, 1917, the United States' Espionage Act was heralded as a tool much-needed to protect military information and public morale following the country's entry into the First World War. Contentious from the moment it was first drafted, the Act's passage through Congress was immune from controversy. While the version eventually enacted was, by many standards, but a shadow of that which had first been proposed, it was not without its ambiguities. These legal nuances would prove crucial to the Act's enforcement, where the judiciary, unable to agree on the limits of its application, largely chose to interpret the Act broadly, thus securing the prosecutions of more than 2,000 alleged dissenters. Irrespective of Congress's intention to simply prevent obstruction of the war effort,

---

1   With many thanks to the editors for their insightful and helpful comments and suggestions. And special thanks to Professor David V. Williams, (now) Professor Emeritus and Honorary Research Fellow at the Faculty of Law, University of Auckland, who was the first to read and review this chapter.

the judiciary felt obliged to safeguard not only the national defense but the confidence of the home front as well. The Espionage Act, their chosen vehicle of patriotic protection (and that which history would later paint as an unprecedented assault on American civil liberties) evolved to become a watershed in the modern fight for the freedom of expression.

Keywords: Espionage Act 1917, Sedition Act 1918, espionage, sedition, anti-war, anti-draft, anti-government, propaganda, freedom of expression, freedom of speech, free speech, World War I, war, wartime, home front, private rights, political rights, civil rights

Lying, Spying, and Right Defying: The Espionage Act of 1917 and the US Wartime Qualification to the Freedom of Speech

When 15-year-old Mollie Steimer arrived at Ellis Island with her household of seven in 1913, she and her family were little more than another group of immigrants fleeing the poverty, violence, and anti-Semitism of tsarist Russia. At 4 ft, 9 in. tall and weighing less than 90 lb, Steimer was but a "mere slip of a girl" (Stone, 2004). Two days after her arrival, she commenced work at a garment factory on New York's Lower East Side. It was dreary and dispiriting work, and she turned to the radical writings of Mikhail Bakunin, August Bebel, and Emma Goldman for comfort. By the time Britain had declared war on Germany in late 1914, Steimer had befriended Goldman and begun to involve herself in trade union activities. By 1917, at 19 years old, Mollie Steimer was a committed anarchist.

At the outbreak of the February Revolution in Russia, Steimer joined a group of fellow revolutionaries publishing surreptitious journals in Yiddish, where they developed Thomas Jefferson's reflection that "[that] government is best which governs least" to "[that] government is best which governs not at all" (Stone, 2004). Upon President Wilson's declaration that the United States had entered into the war, Steimer and her associates found themselves in a political frenzy. Incessantly publishing leaflets which opposed the US intervention in Germany, and which declared the war an imperialist and counter-revolutionary distraction from the worker's struggle, the group called on workers to refuse the draft for military service and to participate in a general strike to protest the government's actions. Word of their seditious intentions soon reached federal officials, who stormed the group's Harlem headquarters and arrested Steimer and four of her comrades. All five were charged under the Espionage Act of 1917 with conspiracy to "publish disloyal material intended to obstruct the war and cause contempt for the government of the United States" (Stone, 2004). On October 10, 1918, the case went to trial; Steimer and her comrades were found guilty and sentenced to 15 years' imprisonment. They appealed, arguing their prosecutions and the Act under which they had been charged were a breach of their First Amendment right to free speech. The case would eventually reach the US Supreme Court.

To pass off the group's constitutional claims as the mere protests of disgruntled rebels is to assume Steimer and her colleagues were alone in questioning the Espionage Act's legal legitimacy in regards to free speech. They were not. When it was first passed in June 1917, the Espionage Act was heralded as a "much-needed tool" to protect military information and public morale following the US entry into the war (Rothe, 2007). The ambiguities of the legislation, however, soon saw the Act used in more than 2,000 prosecutions by the Department of Justice, with its prohibition on publication applied to numerous left-liberal and socialist publications (Manz, 2007). As enforced during the war, it was said to have constituted the "broadest attack on civil liberties in American history" (Manz, 2007).

Introduced into Congress for one reason, passed for another, and enforced for a third, the Espionage Act of 1917 helped to create an environment where expressions of opinion could be punished as conspiratorial lies, where one's neighbor could be arrested on the merest suspicion of being a German spy, and where the constitutional right to freedom of speech could be denied in the interests of national defense.

The Espionage Act

The Act

Officially titled "An Act to punish acts of interference with the foreign relations, the neutrality, and the foreign commerce of the United States, to punish espionage and better enforce the criminal laws of the United States and for other purposes," the Espionage Act was designed to cover a broad scope of offences (Espionage Act of 1917, Pub. L. 64–24, 40 Stat 217, Ch 30, H R 291, 65th Congress; hereafter "Espionage Act 1917, 40 Stat 217"). These ranged from espionage to injuring vessels engaged in foreign commerce, enforcing neutrality, seizing arms and other articles intended for export, counterfeiting the government seal, and the improper use of mails (mails here quite literally meaning all letters, postcards, newspapers, books, publications "conveyed in the mails or delivered from any post office"). This analysis will be primarily concerned with Title I of the Act, entitled "Espionage," and to some extent Title XII, the "Use of Mails," for it was under these two provisions that the majority of wartime prosecutions occurred.

Title I covered those who obtained (a) information at places connected with the national defense; (b) copies of national defense plans; or (c) anything connected with the national defense and did so for the purpose "of obtaining information respecting the national defense with intent or reason to believe that the information to be obtained is to be used to the injury of the United States" (Espionage Act 1917, 40 Stat 217, 217). Whoever violated the terms of this section "in time of war" was liable to punishment by death or imprisonment for up to 30 years.

Title XII of the Act covered "every letter, writing, circular, postal card, picture, print . . . newspaper, pamphlet, book or other publication" which was in violation of the Act's provisions, or which "contained any matter advocating or urging treason, insurrection or forcible resistance to any law of the United States" (Espionage Act 1917, 40 Stat 217, 230). Any printed material which was found to have violated these terms was declared to be "nonmailable matter" and would not be delivered or conveyed via the postal service.

The Passing of the Act

The Espionage Act was officially passed into law on June 15, 1917, almost 10 weeks after it was first introduced into Congress. The version of the Act that was finally adopted was significantly less repressive than the draft originally proposed. Much of Congress's motivations for constructing the offences contained within the Act can be deduced from the congressional debates. The most heated of these centered on three of the Act's most distinctive provisions, what Stone (2004) has termed: (1) the "'disaffection' provision," (2) the "'nonmailability' provision," and (3) the most controversial of these, the "'press censorship' provision." Each will be discussed in turn.

The Disaffection Provision

Had this provision been adopted as it was first proposed, it would have been unlawful for any person, in time of war, to willfully (a) "make or convey false reports and false statements with intent to interfere with the operation or success of the military or naval force of the United States, or promote the success of its enemies"; or (b) "cause or attempt to cause insubordination, disloyalty, mutiny or refusal of duty in the military or naval forces" (Espionage Act 1917, § 3, 40 Stat 217, 219). Criticism of this section was fierce. Twenty-six laypeople of varying social, economic, geographical and educational backgrounds provided testimony as to their opinion(s) on the proposed Act in hearings before the House Committee on the Judiciary between April 9–12, 1917 (H R Rep No 65-65). The disaffection provision featured prominently in their disparagement. Professor Emily Balch of Wellesley College, for example, warned the Committee that an "act like this" would be enforced under war conditions and should be carefully safeguarded to avoid "war hysteria" (*Espionage and Interference with Neutrality Hearing*, 65th Congress, 1917). A jury charged with deciding whether an accused had intended to cause disaffection, she argued, would be "no less liable to hysteria than those higher up" and could easily confuse intention with expectation.

Further statements expressed similar opinions. Each committee member tirelessly repeated that the section would only prosecute or interfere with the right to free speech where the provision was satisfied—i.e., where disaffection, or interference with the "national defense," was deliberately caused. Of those who gave statements, few accepted this position. Gilbert Roe, an attorney who represented the Free Speech League, testified that this provision was an even greater qualification of freedom of speech than the (now repealed) Sedition Act of 1798. That, he argued, had at least recognized the defense of truth; under the disaffection provision, this was no defense, and it appeared that every effort to criticize or discuss the war could be "brought under the ban" (*Espionage and Interference with Neutrality Hearing*). Furthermore, Mr. Roe added, "the greater the truth [of the statement] the greater the disaffection its dissemination [could] cause" (Stone, 2004). Mr. Horace Easton of Syracuse, New York, likewise observed that there "is a distinct vagueness among the members of the committee as to how far this abridgment of speech is going" (*Espionage and Interference with Neutrality Hearing*). Evidently, the House Committee on the Judiciary came to see their point, finding the term "disaffection'" to be too "broad," "elastic," and "indefinite" (55 Cong. Rec. 1530, 1594 (1917)). Difficult to clarify or define further, the phrase "cause or attempt to cause disaffection" was replaced with "cause or attempt to cause insubordination, disloyalty, mutiny or refusal of duty" (ibid; Stone, 2004).

The Nonmailability Provision

This provision would have granted the postmaster general the authority to exclude from the mail any publication or writing which violated "the provisions of the act" or which was otherwise "of a treasonable or anarchistic character" (Stone, 2004). In contrast to the disaffection provision, the nonmailability provision was most strongly opposed by members of Congress, who protested granting the postmaster general the authority to exclude even political mail (Stone, 2004). Representative Meyer London of New York, for example, claimed the provision was a "menace to freedom" (55 Cong. Rec. 1756, 1806 (1917)). Senator Charles Thomas of Colorado argued the provision would produce a "far greater evil than the evil which is sought to be prevented," while Senator William Borah declared "the more you dam up the stream the more liable you are to have a flood when it breaks over" (ibid; Stone, 2004).

The congressional members' concerns were tied to two specific aspects of this provision. They first objected to the provision as a whole, arguing the provision gave the postmaster general the power to exclude both legitimate and illegitimate publications. Such authority could become a slippery slope to "democracy gone mad" (Stone, 2004). Opposition was also directed at the use of the words "treasonable" and "anarchistic" in the provision. These were felt to be incredibly vague and capable of manipulation at the whim of whoever "happens to be high in the Post Office Department" (Stone, 2004). After a spirited debate from its members, Congress agreed to amend the provision. The phrase "treasonable or anarchistic character" was replaced with the narrower, more determinate phrase "containing any matter advocating or urging treason, insurrection or forcible resistance to any law of the United States" (Stone, 2004). Furthermore, only those statements which demonstrated express advocacy of "treason, insurrection or forcible resistance" could be excluded from the mail (Stone, 2004). Given that Congress appeared to take into account the constitutional and policy objections to granting too broad a discretion to the postmaster general, this qualification became an important insight into Congress's intended meaning of this provision later in the war.

The Press Censorship Provision

This third provision, despite its eventual congressional defeat, is nonetheless worth noting, for it proved the most controversial of all the Act's proposed provisions. If passed, the provision would have declared it unlawful for any person "in time of war" to publish any information that the president had declared to be "of such character that it is or might be useful to the enemy" (Stone, 2004). Nothing in this section, furthermore, could be "construed to limit or restrict any discussion, comment, or criticism of the acts or policies of the Government" (Stone, 2004). Proposed by the Wilson administration, this provision generated a "firestorm of protest," most especially from the press, who claimed this effectively gave the president the final authority to determine what information could or could not be published concerning the conduct of the war (Stone, 2004).

The American Newspaper Publishers Association, for example, declared that this provision struck "at the fundamental rights of the people, not only assailing their freedom of speech, but also seeking to deprive them of the means of forming an intelligent opinion (Stone, 2004). In war especially, the Association added, the press should be "free, vigilant and unfettered" (Stone, 2004). The *New York Times* attacked the provision as "high-handed" and "Prussian," observing that those who criticized or highlighted defects in policies "with the honest purpose of promoting remedial action . . . is not a public enemy" (Stone, 2004). The *Milwaukee News* characterized it as a "glaring attempt . . . to muzzle the press," while the *Philadelphia Evening Telegraph* professed that there should be "no power in this country . . . to control the voice of the press, or of the people, in honest judgment of the acts of public servants" (Stone, 2004).

Congressional debate over the provision was fiercely heated. Some, such as Representative Edwin Webb of North Carolina, claimed the press needed reminding that "in time of war, while men are giving up their sons [and] people are giving up their money," the press should be willing to give up its right to publish that which the president considers "hurtful to the United States and helpful to the enemy" (*Espionage and Interference with Neutrality Hearing*). Senator Lee Overman, also of North Carolina, likewise argued that "the good of society is superior to the right of the press to publish what it pleases"; if the newspapers were a hindrance to the war effort, their curtailment would not be unconstitutional (*Espionage and Interference with Neutrality Hearing*). Representative Andrew J. Volstead of Minnesota cut straight to the heart of the matter, brusquely asking how the nation would feel if American soldiers were "sent to the bottom of the sea as a result of information" published because Congress had failed to enact this provision (*Espionage and Interference with Neutrality Hearing*).

Representative Walter Chandler of New York, however, objected to the provision because "it is an abridgement of the rights and privileges of the American people to have light thrown on certain things through the Press" (55 Cong. Rec. 1756, 1780 (1917)). Some restriction upon the press, he argued, was not inappropriate, but this should not "put in the hands of a single citizen, be he the President of the United States, the power to say to 100,000 people that which they should know they cannot know" (ibid). Representative Moses P. Kinkaid of Nebraska agreed, arguing that the "scope of this clause [being a qualification on what information is able to be published] is covered by other sections of this bill" (55 Cong. Rec. 3632, 3632 (1917)). Representative Martin B. Madden of Illinois, however, made perhaps the most compelling of these arguments, contending that while America was fighting to establish "the democracy of the world, we ought not to do the thing that will establish autocracy in America" (55 Cong. Rec. 1756, 1800 (1917)).

When it began to appear that the press censorship provision would be defeated, President Wilson appealed to Congress directly in May 1917, writing to Representative Edwin Webb, Chair of the Committee on the Judiciary, that he had

> every confidence that the great majority of the newspapers of this country will observe a patriotic reticence about everything whose publications could be of injury but in every country there are some persons who cannot be relied upon and whose interests or desires will lead to actions on their part highly dangerous to the nation in the midst of war.[2] (Manz, 2007)

Unfortunately for Wilson, Congress was unmoved, for the provision was defeated in the House of Representatives by 184 votes to 144, with 36 Democrats joining the Republican opposition. Following this, any further considerations regarding the censorship of the press were effectively settled for the remainder of the war.

---

2   Original held at Woodrow Wilson Presidential Library, Staunton, VA, No. WWP20621.

After almost two and a half arduous months of political debate, Congress finally enacted the Espionage Act of 1917. The legislation passed was not a broadside attack on all criticism of the war. Rather, it was a carefully considered and well-debated enactment designed, in the words of Assistant Attorney General John Lord O'Brian, to "protect the process of raising and maintaining our armed forces from the dangers of disloyal propaganda" (Manz, 2007). While Congress and the public seemed generally happy with the final version of the Act, the Wilson administration was largely dissatisfied. Attorney General Thomas Gregory complained to the American Bar Association that "most of the teeth which we tried to put in were taken out" (Manz, 2007). The Act, however, clearly still retained some prosecutorial bite, as its enforcement during the course of the war would demonstrate. When Gregory later declared that war dissenters "need expect none [mercy or leniency] from an outraged people and an avenging government," it was clear any executive restraint, in enforcement of the Act, would be little more than wishful thinking (Stone, 2004). Indeed, the perceived deficiencies in the Espionage Act were addressed the next year, when Congress amended Title I by passing the Sedition Act of 1918, which extended Title I to cover those who "wilfully utter, print, write or publish any disloyal, profane, scurrilous or abusive language about the form of the government of the United States . . . or any language intended to incite, provoke, or encourage resistance to the United States or to promote the cause of its enemies" (Sedition Act of 1918, Pub. L. 65–150, 1 Stat 596, Ch 74, H R 8753, 65th Congress). The Sedition Act, which was largely deployed as an extension of the Espionage Act, was justified on the grounds that the 1917 Act was the cause of the rumored "vigilante violence" towards disloyal dissenters, where disgruntled members of the public apparently exercised extralegal sanctions to compensate for the perceived inability of the Act to do the same (Manz, 2007).

Enforcement During the War

Until the passing of the Sedition Act, it was clear the Espionage Act was intended to restrict only expressions that *willfully* caused or attempted to cause "insubordination, disloyalty, mutiny or refusal of duty" in the military and naval forces or if "the recruiting and enlistment service" was obstructed (Stone, 2003). Although the precise line between speech which fell within these prohibitions and speech which did not was far from clear, there is little doubt that, in 1917, Congress intended for such a line to exist. Few members of the judiciary, however, appeared to agree. Upon its enforcement, it was soon apparent that a discrepancy between legislative intent and apparent effect had arisen. As the war progressed and patriotic fervor "whipped into fever pitch," the federal courts increasingly claimed that persistence of prosecution was necessary "in order to suppress a broad [and ever-increasing] range of political dissent" (Stone, 2004).

A few judges, it must be noted, did stand fast in their refusal to be swept away in the patriotic fanaticism. Federal District Judge George Bourquin of Montana, for example, heard the case of Ves Hall, who was alleged to have declared, multiple times, that he would "flee to avoid going to the war," that "Germany would whip the United States," and that "the President was a Wall Street tool" (*United States v. Hall*, 1919). Bourquin directed a verdict of acquittal, finding that Hall's comments, which were made in a Montana village of "some sixty people, sixty miles from the railway" and "hundreds of miles" from any soldiers or sailors, could not justifiably or reasonably be held to have been made with an intention to interfere with the "operations or success of the military" (ibid). Furthermore, Bourquin argued, the Espionage Act had been enacted to suppress not general "criticism, denunciation . . . slander [or] gossip" but only those acts denounced as crimes within the confines of the Act.

The most important judicial decision which held against a broad interpretation of the Espionage Act was *Masses Publishing Co. v. Patten* (1917), heard by Judge Learned Hand. The *Masses* was a monthly "revolutionary" journal that regularly featured the writings of radical liberals such as Max Eastman, Vachel Lindsay, Bertrand Russell, and Carl Sandburg (Stone, 2003). Described as "iconoclastic, impertinent and confrontational," the *Masses* contained lively social satire, political criticism, and intellectual commentary, and it was claimed by contemporary readers to be "not so much a publishing venture as a movement . . . a way of life" (Stone, 2003). In the summer of 1917, however, Postmaster General Albert Burleson ordered the August issue of the *Masses* to be excluded from the mail, exercising his authority under Title XII of the Espionage Act.

Burleson argued that four cartoons and four pieces of text violated the Act by willfully causing or attempting to cause "insubordination, disloyalty, mutiny or refusal of duty in the military or naval forces" and by thus obstructing "the recruiting or enlistment service of the United States" (*Masses Publishing Co. v. Patten*, 1917). One of these "inflammatory" pieces of text was an alleged tribute to Emma Goldman and Alexander Berkman, two of the "most notorious anarchists in the United States" (*New York Times*, 16 June 1917) who were imprisoned for anti-draft conspiracy. (Goldman and Berkman had been convicted, under the Espionage Act, of "overt interference with the nation's war programme" through their anti-war publications, which had allegedly induced "various men of conscript age not to comply with the provisions of the selective draft law" (ibid.).) The poem in question, written anonymously, included this verse:

Emma Goldman and Alexander Berkman
Are in prison tonight,
But they have made themselves elemental forces,

Like the water that climbs down the rocks;
Like the wind in the leaves;
Like the gentle night that holds us;
They are working on our destinies;
They are forging the love of the nations…
Tonight they live in prison. (cited in *Masses Publishing Co. v. Patten*, 1917)

The *Masses* sought an injunction to prevent the local postmaster from refusing to accept the August issue for mailing. Judge Hand granted the injunction, arguing that to read the word "cause" so broadly would involve "necessarily as a consequence the suppression of all hostile criticism, and of all opinion except what encouraged and supported the existing policies" (*Masses Publishing Co. v. Patten*, 1917). Such an approach would, Hand contended, "contradict the normal assumption of democratic government" (*Masses Publishing Co. v. Patten*, 1917). Furthermore, even assuming Congress had the power or intention to "repress such opinion," the exercise of that power, Hand challenged, would be "so contrary to the use and wont of our people" that only the "clearest expression of such a power" would justify the conclusion that it was intended (*Masses Publishing Co. v. Patten*, 1917). In these circumstances, he concluded, the "actual language" of the statute did not require such an interpretation, nor, more importantly, was such an interpretation reasonable in the *Masses* case (*Masses Publishing Co. v. Patten*, 1917).

Hand also gave a second reason for his interpretation. Observing that Congress had expressly chosen to eliminate the press censorship provision, narrow the disaffection provision, and restrict the nonmailability provision, one could reasonably conclude, he held, that Congress had not intended for all criticism that "could arouse discontent and disaffection among the people" to be criminalized (Stone, 2004). Had Congress wanted to prohibit all such expression, it "could surely have said so" (Stone, 2004). What Congress had said, in the wording of the Espionage Act, was that free expression could be restricted only where, and if, it *willfully* undermined or attempted to undermine "the effectiveness of the nation's armed forces" (Stone, 2004). What was required, in other words, was express advocacy of such critical expression. Finding no such willfulness on the part of the *Masses*, Hand dismissed the espionage charges and granted the injunction.

Few other members of the wartime judiciary, however, followed the lead of Bourquin, Hand, or their likeminded peers. Instead, most judges appeared determined to read the Espionage Act liberally, imposing severe sentences on those charged with disloyalty, dissent, or sedition. John Lord O'Brian, head of the War Emergency Division of the Department of Justice, later observed that, in the wartime atmosphere of "excessive passion, patriotism and clamour," the laws affecting free speech received the "severest test thus far placed upon them in our history" (Stone, 2004).

The prevailing approach taken by the judicial majority is best illustrated by *Shaffer v. United States* (1919), a US Court of Appeals for the Ninth Circuit case. Here, the defendant was charged with possessing and mailing copies of a book, *The Finished Mystery* (whose author was unknown), which contained this passage:

> a certain delusion which is best described by the word patriotism, but which is in reality murder, the spirit of the very devil. If you say it is a war of defense against wanton and intolerable aggression, I must reply that . . . it has yet to be proved that Germany has any intention of [*sic*] desire of attacking us . . . The war itself is wrong. Its prosecution will be a crime. (cited in *Shaffer v. United States*, 1919)

The Court of Appeals, which upheld Shaffer's prosecution, reasoned that the question which determined whether the Act had been breached was "whether the natural and probable tendency and effect of the words . . . are such as are calculated to produce the result condemned by the statute" (*Shaffer v. United States*, 1919). This "bad tendency" check thus became the judicial litmus test by which seditious conduct was to be measured, and it would quickly be adopted by almost every US court. The fears of those who had testified before the House Committee on the Judiciary during the passing of the Act, that the judges and juries sitting on espionage cases during the war could become incontrovertibly "swayed by wartime hysteria," appeared to have begun to ring true (Stone, 2004).

The case of Rose Pastor Stokes was one such example of this judicial hysteria. A Russian immigrant, Stokes was editor of the socialist *Jewish Daily News*. She was convicted under the Espionage Act for saying "I am for the people and the government is for the profiteers" during a speech to the Women's Dining Club of Kansas City; this was later published in the *Kansas City Star* (*United States v. Stokes*, 1918). Although there were no men, let alone soldiers, in Stokes's audience, the government claimed she had violated the Espionage Act because "our armies . . . can operate and succeed only so far as they are supported and maintained by the folks at home" (Stone, 2004). The court accepted this argument, holding that Stokes's speech had the potential to "chill enthusiasm, extinguish confidence and retard [the] cooperation [of] mothers, sisters, and sweethearts" (Stone, 2004). She was sentenced to 10 years in prison. Stokes's conviction was reversed by the Supreme Court in 1920 (Kennedy, 1999). She was never retried.

Thirty German Americans in South Dakota, likewise, were convicted for sending the government a petition requesting reforms in the selective service procedure (Stone, 2004). "Threatening" to vote the local governor out of office if he did not meet their demands, the group was found guilty of willfully obstructing the recruiting and enlistment service (Stone, 2004). Robert Goldstein, similarly, was convicted for producing and later exhibiting a motion picture about the American Revolution. *The Spirit of '76* included a scene which depicted the Wyoming Valley Massacre, where British soldiers murdered women and children. The government argued that this was an attempt to "promote insubordination" because it was portraying Britain, America's ally in the current war against Germany, in a negative light (*United States v. Motion Picture Film "The Spirit of '76,"* 1917). The judge overseeing the trial agreed, holding that war was "no time" for "those things that may have the tendency or effect of sowing . . . animosity or want of confidence between us and our allies" (*United States v. Motion Picture Film "The Spirit of '76,"* 1917). Goldstein was sentenced to 10 years in prison, although he would only end up serving 3—President Wilson later commuted his sentence (*New York Times,* 1921).

As assistant attorney general, John Lord O'Brian would observe in 1919 that, "with a degree of unanimity which [was] extraordinary," the courts during the war had insisted that Title I of the Espionage Act was "broadly intended by Congress to include every form of activity, by speech or conduct, which was [willfully] intended by direct or indirect means to obstruct the work of raising and maintaining the National Armies" (Manz, 2007). In the hands of the federal judiciary, the Act became an "efficient tool for the blanket suppression" of all disloyal utterances (Stone, 2004). None of the accused in the aforementioned cases had expressly advocated insubordination, disloyalty, refusal of duty, obstruction of the recruiting and enlistment services, or any other seditious activity as defined in the Act. Each had, however, made the mistake of questioning the legality, morality, or rationality of the war, and it was the "natural and probable tendency" of this which the courts felt would—or could—obstruct the war effort (Stone,

2004). It should come as no surprise to note that opposition to the war became synonymous with conviction under the Espionage Act.

Although the war officially ended on November 11, 1918, prosecutions under the Act continued for some time. Most of these were appeals attempting to ascend the judicial hierarchy on claims that the Act was unconstitutional. The first of these to reach the Supreme Court was *Schenck v. United States* (1919), which was decided in March 1919. In *Schenck*, the defendants had been charged with conspiring to obstruct the recruiting and enlistment service by circulating a pamphlet to men who had been called for military service. The pamphlet argued that the draft was unconstitutional, that conscription was "little better than a conflict," was a "monstrous wrong," and was designed to further the interests of Wall Street (*Schenck v. United States*, 1919). It urged readers not to "submit to intimidation" and warned that "if you do not assert and support your rights, you are helping to deny or disparage rights which it is the solemn duty of all citizens and residents of the United States to retain" (*Schenck v. United States*, 1919).

Despite only a few inductees actually receiving the pamphlet, and none providing evidence that they had been influenced by it, Schenck and his codefendants were nevertheless found guilty because the "natural and probable tendency" of the pamphlet was to "dampen the willingness of men to serve in the armed forces" and because the jury could reasonably infer that the defendants had "intended to cause the natural and probable consequence of their actions" (Stone, 2004). The Supreme Court upheld the conviction, asserting that when a nation is at war, "many things might be said in time of peace which are such a hindrance to its efforts that their utterance will not be endured so long as men fight, and no Court could regard them as protected by any constitutional right" (*Schenck v. United States*, 1919). The question in every case, the court contended, was "whether the words used are used in such circumstances and are of such a nature as to create a clear and present danger that they will bring about the substantive evils that Congress has a right to prevent" (*Schenck v. United States*, 1919). This, the court maintained, was itself a "question of proximity and degree" (*Schenck v. United States*, 1919).

The "clear and present danger" test endorsed by the Supreme Court in *Schenck* was little more than a reformulation of the "natural and probable tendency" standard disseminated by the courts below. Given the decided lack of deviation on this judicial theme, it is unsurprising that two other appeals to the Supreme Court, heard within weeks of *Schenck*, were decided the same way. Jacob Frohwerk, a copy editor for the German-language newspaper *Missouri Staats Zeitung*, was convicted of breaching the Espionage Act for publishing a series of articles over a 6-month period in 1917. These included statements that declared "it a monumental and inexcusable mistake to send our soldiers to France," that US participation in the war "appears to be outright murder without serving anything practical," and that the war had been entered into "so that a few men and corporations might amass unprecedented fortunes while we sold our honor, our soul" (*Frohwerk v. United States*, 1917). While it did not appear that there was any special effort to reach men who were subject to the draft, the Supreme Court found it "impossible to say that it might not have been found that the circulation of the paper was in quarters where a little breath would be enough to kindle a flame" (*Frohwerk v. United States*, 1917). Frohwerk's conviction was upheld.

A similar verdict was reached in the appeal of Eugene Debs. An infamous member of the Socialist Party of America and a major national figure, Debs strongly opposed both conscription and the American intervention in the war. On June 16th, 1918, following a visit to three Socialist party members imprisoned in Ohio for violating the Espionage Act, Debs gave the following speech to a crowd of 1,200:

> It is extremely dangerous to exercise the constitutional right of free speech in a country fighting to make democracy safe in the world. I realize that, in speaking to you this afternoon, there are certain limitations placed upon the right of free speech. I must be exceedingly careful, prudent, as to what I say, and even more careful and prudent as to how I say it. I may not be able to say all I think, but I am not going to say anything that I do not think... They tell us that we live in a great free republic; that our institutions are democratic; that we are a free and self-governing people. This is too much, even for a joke. But it is not a subject for levity; it is an exceedingly serious matter. (*Debs v. United States*, 1919)

For this, Debs was arrested, convicted, and sentenced to a prison term of 10 years for obstructing the recruiting and enlistment services. Contending that "the delivery of a speech in such words and such circumstances that the probable effect will be to prevent recruiting," the Supreme Court upheld the conviction.

In each of these cases the judgment of the Supreme Court was delivered by Justice Oliver Holmes. Holmes was a close friend of Learned Hand, and shortly after the Supreme Court decision in *Debs*, Holmes received a letter from Hand questioning the court's interpretation of the Act (Polenberg, 1987). Hand argued that "the thing against which the statute aims is positive impediments to raising an army," and speech which violated the Act only did so "when the words were directly an incitement" (Polenberg, 1987). Holmes replied to Hand in early April, claiming, "I don't get your point," adding that he could not see any difference between Hand's "direct incitement" test and his own clear and present danger standard (Polenberg, 1987). The following month, an article by Ernst Freud appeared in *The New Republic*, where Freud voiced his opinion that the clear and present danger test was precariously vague and that juries, swayed by hysteria, would be "overly anxious to convict dissenters" (Polenberg, 1987). The peril to the national cause as a result of toleration of adverse opinion, Freud wrote, was "in any event . . . slight compared with the permanent danger of intolerance to free institutions" (Polenberg, 1987). Although Holmes reacted to Freud's criticism as he did to Hand's, he would later admit he had already begun to harbor nagging doubts regarding the espionage convictions the Supreme Court had most recently reached.

It was in this "unsettled frame of mind" that Holmes discovered Zechariah Chafee's article "Freedom of Speech in Wartime" in the June 1919 issue of the *Harvard Law Review* (Polenberg, 1987). Chafee argued that a "provision like the First Amendment to the Federal Constitution" was "much more than an order to Congress not to cross the boundary which marks the extreme limits of lawful suppression" (Chafee, 1919). It was, he contended, a "declaration of national policy in favor of the public discussion of all public questions" which should "influence the judges in their construction of valid speech statutes, and the prosecuting attorneys who control their enforcement" (Chafee, 1919). Chafee was not asserting that the right to speech should have no limitations; rather, he was emphasizing that the classification of speech as lawful or unlawful involves "the balancing against each other of two very important social interests, in public safety and in the search for truth" (Chafee, 1919). Every reasonable attempt, he stressed, "should be made to maintain both interests unimpaired," with free speech only sacrificed when "the interest in public safety is really imperiled" (Chafee, 1919). Having considered Chafee's arguments seriously, Holmes had by autumn begun to rethink the issue of free speech. The first case the Supreme Court heard following Holmes's ideological change of heart was *Abrams v. United States* (1920).

The defendants in *Abrams* were Mollie Steimer and her comrades. As with all appeals the Supreme Court heard in the years immediately following the war, Steimer and her colleagues were claiming their espionage convictions were a breach of their First Amendment rights. Predictably, the Supreme Court affirmed the convictions, noting that "this contention is sufficiently discussed and is definitely negatived" in *Schenck, Frohwerk,* and *Debs* (*Abrams v. United States*, 1920). Unexpectedly, however, Justice Holmes (joined by Justice Brandeis) dissented. Holmes began his dissent by noting that he did not see "any reason to doubt that . . . *Schenck, Frohwerk,* and *Debs* were rightly decided" (*Abrams v. United States*, 1920). By the same reasoning, however, he declared,

> The United States constitutionality may punish speech that produces or is intended to produce a clear and imminent danger that it will bring about forthwith certain substantive evils . . . the power undoubtedly is greater in time of war than in time of peace because war opens dangers that do not exist at other times. . . . It is only the present danger of immediate evil or an intent to bring it about that warrants Congress in setting a limit to the expression of opinion where private rights are not concerned. (*Abrams v. United States*, 1920)

In the case of the *Abrams* defendants, he argued, no one would presume that the "surreptitious publishing of a silly leaflet by an unknown man, without more, would present any immediate danger that its opinions would hinder the success of the government [or] have any appreciable tendency to do so" (*Abrams v. United States*, 1920). In Holmes's opinion, Steimer and her colleagues justifiably deserved to have been acquitted.

Conclusion: Post-Espionage

Following the signing of the Armistice in November 1918, which signaled the official end of the war, the question of amnesty for those currently imprisoned for convictions under the Espionage Act began to be contemplated, for the threats of foreign influence which had been so feared were no longer of significant concern. Before leaving office in 1919, Attorney General Gregory, on the advice of Assistant Attorney General O'Brian, recommended to President Wilson the release or reduction in sentence of 200 espionage prisoners (Stone, 2004). These convictions, Gregory had explained, were the result of the "intense patriotism" of the jurors, and the sentences received were "out of proportion" to the offense (Stone, 2004). It was now clear, he contended, that injustices had been done—these people were "in no sense political prisoners" but simply "criminals who sided against their country" during a time of immense political pressure and xenophobic fear (Stone, 2004). The president accepted these recommendations. Over the following years, sustained efforts, both publicly and privately, were made to secure the release of the Act's remaining prisoners. On Christmas Day, 1921, Wilson's successor, President Warren G. Harding, pardoned Eugene Debs and 24 other espionage convicts (Stone, 2004). In December 1923, President Coolidge ordered the release of all remaining prisoners (Stone, 2004). 10 years later, in 1933, President Franklin Delano Roosevelt granted amnesty to all individuals convicted under the Espionage and Seditions Acts, restoring their full political and civil rights (Stone, 2004). On December 13, 1920, the Sedition Act of 1918 was quietly repealed by Congress. The Espionage Act, however, remained in force.

Since its enactment in 1917, the Act has maintained a well-deserved place in the political, social, legal, and academic debates concerned with the wartime qualification of the freedom of speech. Passed by a Congress who wanted only to prevent obstruction with the war effort and administered by a judiciary who felt obligated to safeguard the national defense and neighborhood morale, the Espionage Act of 1917 became a watershed in the fight for freedom of expression, imprisoning those who lied, prosecuting those who spied, and allowing rights of expression to be denied. As for Mollie Steimer, when she and her four seditious co-conspirators were released on bail awaiting their appeal to the Supreme Court, she immediately resumed her political activities and was consequently rearrested, imprisoned, and threatened with deportation (Stone, 2004). Following the upholding of her conviction in *Abrams*, Steimer was deported to Russia on April 30, 1920. Dedicating her entire life to the worker's cause, she remained constantly on the move, evading authorities determined to imprison her for her anarchistic activities. Deported from Russia to Germany in 1927, Steimer later fled to Mexico in response to the rise of the National Socialist Party (Stone, 2004). She remained in Mexico for the rest of her life, where she died on July 23, 1980, aged 82 years old, unmarried, with no children, and the only one of the original five defendants in the *Abrams* case to have remained a committed anarchist.

References

54 Cong Rec 3142 (8 February 1917).

54 Cong Rec 3828 (16 February 1917).

54 Cong Rec 3913 (17 February 1917).

54 Cong Rec 3982 (19 February 1917).

54 Cong Rec 3994 (19 February 1917).

54 Cong Rec 4077 (19 February 1917).

54 Cong Rec 4136 (20 February 1917).

55 Cong Rec 760 (18 April 1917).

55 Cong Rec 833 (19 April 1917).
55 Cong Rec 879 (20 April 1917).
55 Cong Rec 907 (21 April 1917).
55 Cong Rec 914 (21 April 1917).
55 Cong Rec 1165 (26 April 1917).
55 Cong Rec 1530 (30 April 1917).
55 Cong Rec 1657 (1 May 1917).
55 Cong Rec 1685 (2 May 1917).
55 Cong Rec 1702 (2 May 1917).
55 Cong Rec 1756 (3 May 1917).
55 Cong Rec 1823 (4 May 1917).
55 Cong Rec 1848 (4 May 1917).
55 Cong Rec 1886 (4 May 1917).
55 Cong Rec 1892 (4 May 1917).
55 Cong Rec 1894 (5 May 1917).
55 Cong Rec 1909 (5 May 1917).
55 Cong Rec 1947 (5 May 1917).
55 Cong Rec 1968 (7 May 1917).
55 Cong Rec 2020 (8 May 1917).
55 Cong Rec 2022 (8 May 1917).
55 Cong Rec 2133 (10 May 1917).
55 Cong Rec 2146 (10 May 1917).
55 Cong Rec 2202 (11 May 1917).
55 Cong Rec 2224 (11 May 1917).
55 Cong Rec 2250 (12 May 1917).
55 Cong Rec 2336 (14 May 1917).
55 Cong Rec 3104 (25 May 1917).
55 Cong Rec 3238 (29 May 1917).
55 Cong Rec 3257 (29 May 1917).

55 Cong Rec 3330 (31 May 1917).

55 Cong Rec 3363 (31 May 1917).

55 Cong Rec 3497 (4 June 1917).

55 Cong Rec 3499 (5 June 1917).

55 Cong Rec 3508 (6 June 1917).

55 Cong Rec 3557 (7 June 1917).

55 Cong Rec 3632 (8 June 1917).

55 Cong Rec 3638 (8 June 1917).

55 Cong Rec 3703 (11 June 1917).

55 Cong Rec 3769 (12 June 1917).

Abrams v. *United* States (1919), 250 US 616.

Axelrod, A. (2007). *Political history of America's wars.* Washington, DC: CQ Press.

Capozzola, C.J. (2008). *Uncle Sam wants you: World War I and the making of the modern American citizen.* Oxford, UK: Oxford University Press.

Cassidy, G.A. (1932). Proposed amendments to the Federal Espionage Act of 1917. *Georgetown Law Journal 21*, 339–344.

Chafee, Z. (1919). Freedom of speech in war time. *Harvard Law Review 32*, 932–973.

Debs v. *United States* (1919), 249 US 211.

Edgar, H., & Schmidt, B.C. (1973). The espionage statutes and publication of defense information. *Columbia Law Review 73*(5), 929–1087.

Epstein, R.D. (2006). Balancing national security and free-speech rights: Why Congress should revise the Espionage Act. *Commlaw Conspectus, 15,* 483–516.

Ericson, T.L. (2005). Building our own 'Iron Curtain': The emergence of secrecy in American government. *American Archivist, 68,* 18–52.

Espionage Act of 1917, Pub. L. 64–24, 40 Stat 217, Ch 30, H R 291, 65th Congress.Espionage and Interference with Neutrality: Hearing on H.R.

291 Before the H. Comm. of the Judiciary, 65th Cong. (1917).

Frohwerk v. United States (1917), 249 US 204.

H R Rep No 64-1449 (1917).

H R Rep No 64-1591 (1917).

H R Rep No 65-30 (1917).

H R Rep No 65-65 (1917) (Conf Rep).

H R Rep No 65-69 (1917) (Conf Rep).

Herbeck, D.A. (1987). Fair play did not permit excess: A critical review of the histories of the Espionage Act of 1917. *Free Speech Yearbook, 26*(1), 11–27.

Hester, J.L. (2001). The Espionage Act and today's 'high-tech terrorist'. *North Carolina Journal of Law & Technology, 12*, 177–197.

Hilton, O.A. (1947). Public opinion and civil liberties in wartime: 1917–1919. *Southwestern Science Quarterly, 28*(3), 201–224.

Kennedy, K. (1999). Disloyal Mothers and Scurrilous Citizens: Women and Subversion During World War I. Bloomington, IN: Indiana University Press.

Kimball, D. (1920). The Espionage Act and the limits of legal toleration. *Harvard Law Review*, 442–449.

Kittrie, N.N., & Wedlock, E.D. (1998). *The tree of liberty: A documentary history of rebellion and political crime in America (2nd ed.)*. Baltimore, MD: John Hopkins University Press, Baltimore, 1998.

Manz, W.H. (2007). *Civil Liberties in wartime: Legislative histories of the Espionage Act of 1917 and the Sedition Act of 1918*. W.H. Manz (Ed.). Buffalo, NY: WS Hein.

Masses Publishing Co. v. Patten (1917), 244 F Supp 535 (SD NY).

New York Times, 'Emma Goldman and A. Berkman Behind the Bars' June 16th, 1917.

New York Times, 'Revive "Spirit of '76," film barred in 1917', July 14th,

1921.

Pierce v. *United States* (1920), 252 US 239.

Polenberg, R. (1987). *Fighting faiths: The Abrams case, the Supreme Court, and free speech.* New York, NY: Penguin Books.

Posner, E. (2007). The war on speech in the war of terror: An examination of the Espionage Act applied to modern First Amendment doctrine. *Cardozo Arts & Entertainment Law Journal, 25,* 717–746.

Rabban, D.M. (1980). The First Amendment in its forgotten years. *Yale Law Journal, 90,* 514–596.

Revision and Strengthening of Espionage, Neutrality, Passport and Shipping Regulations: Hearing on H.R. 291 Before the H. Comm. of the Judiciary, 65th Cong. (1917).

Rothe, D. L. (2007). *Battleground: Criminal justice.* G. Barak (Ed.). Westport, CT: Greenwood Press.

Shaffer v. *United States* (1919), 255 F 886 (9th Cir).

Schenck v. *United States* (1919), 249 US 47.

Sedition Act of 1918, Pub. L. 65–150, 1 Stat 596, Ch 74, H R 8753, 65th Congress.

Smith, C.R. (2011). *Silencing the opposition: How the U.S. Government suppressed freedom of expression during major crises (2nd ed.).* Albany, NY: State University of New York Press.

Stone, G.R. (2002). The origins of the 'bad tendency' test: Free speech in wartime. *Supreme Court Review,* 411–453.

Stone, G.R. (2003). Judge Learned Hand and the Espionage Act of 1917: A mystery unraveled. *University of Chicago Law Review, 70,* 335–358.

Stone, G.R. (2004). *Perilous times: Free speech in wartime.* New York: WW Norton & Co.

Taft, H.W. (1921). Freedom of speech and the Espionage Act. *American Law Review, 55,* 696–721.

Thomas, W.H. (2008). *Unsafe for democracy: World War I and the U.S. Justice Department's covert campaign to suppress dissent.* Madison, WI: University of Wisconsin Press.

United States v. *Hall* (1919), 249 F 150 (DC MO).

United States v. *Motion Picture Film "The Spirit of '76"* (1917), 252 F 946 (SD CA).

United States v. *Schutte* (1918), 252 F 212 (DC ND).

United States v. *Stokes* (1918), unreported; on appeal, (1920), 264 F 18 (8th Cir).

Wigmore, J.H. (1920). Abrams v. U.S.: Freedom of speech and freedom of thuggery in war-time and peace-time. *Illinois Law Review, 14*(8), 539–561.

6

League of Nations Debate: Strategic Preferences of President Woodrow Wilson and Senator Henry Cabot Lodge

Seyed Hamidreza Serri[1]
University of North Georgia

Abstract

This chapter argues that the disagreements in the United States over joining the League of Nations and the subsequent refusal of the United States to join the League can be explained by differences in strategic cultures of supporters and opponents of the League in the US. It studies the strategic cultures of President Woodrow Wilson and Senator Henry Cabot Lodge as representatives of supporters and opponents of joining the League. It also proposes Operational Code Analysis as a method for extracting strategic cultures.

Keywords: Woodrow Wilson, United Nations, League of Nations, Operational Code Analysis

---

1 Special thanks to Dr. Stephen Walker for his invaluable inputs. Also, special thanks to Dr. Michael Young and the Social Science Automation, Inc., for permission to use the Profiler Plus software.

Introduction

President Woodrow Wilson championed the idea of the League of Nations in his Fourteen Points address as follows: "A general association of nations must be formed under specific covenants for the purpose of affording mutual guarantees of political independence and territorial integrity to great and small states alike" (Wilson, 1918). Following President Wilson's Fourteen Points, and during the Paris Peace Conference, the United States strongly pushed for the creation of a "general association of nations" (Wilson, 1918). As a result, the Paris Peace Conference voted for establishing the League of Nations. However, under the leadership and direction of the US Senator Henry Cabot Lodge (R-MA), who was the majority leader and chair of the Senate Committee on Foreign Relations, the Senate rejected the Treaty of Versailles. Consequently, the United States did not join the League of Nations. Why did the Senate reject the Versailles Treaty?

In this chapter, the answer to the question posed above is that the rejection of the League of Nations can be explained by the different strategic cultures of President Woodrow Wilson and Senator Henry Cabot Lodge and their different preferences for political outcomes. To achieve this goal, I assess three hypotheses: (1) Wilson and Lodge's strategic cultures were different; (2) relative to Wilson, Lodge's strategic culture presented a more hostile worldview and higher propensity for conflict; and (3) the deadlock of the League of Nations debate can be explained by Wilson and Lodge's different strategic cultures. I evaluate these three hypotheses in five sections. The first section offers a short history of the debate in the United States about the League of Nations; the second, an explanation of how the operational code method can extract microfoundations of strategic culture; the third, an evaluation of Hypotheses 1 and 2; and the fourth, an evaluation of Hypothesis 3. In the fifth section, I present my conclusion.

The League Fight

On July 10, 1919, and for the first time since 1789, a president hand-delivered a treaty to the US Senate (The United States Senate, n.d.). That treaty was the Treaty of Versailles, which President Wilson had personally negotiated during the Paris Peace Conference. To promote international security and cooperation, Part I of the Treaty of Versailles was dedicated to the Covenant of the League of Nations, which asked the contracting parties to follow international law, and not resort to use of force as the rule of conduct among themselves (The Versailles Treaty, 1919). While the idea of creating an international organization to promote peace and security seems a constructive idea, the decision of whether or not to join the League of Nations caused a spectacular debate in the United States (World Peace Foundation, 1919). Senator Lodge referred to this debate as "one of the most remarkable, if not the most remarkable, which had ever occurred in the Senate of the United States" (Lodge, 1925, p. 178).

In the Senate, the "League fight" created three groups (Cooper, 2010). One group, led by Senator Gilbert Hitchcock, strongly supported the League of Nations and the ideas behind its creation. The second group comprised those who supported joining the League of Nations but had reservations about it. The leader of this group, known as the Reservationists, was Senator Henry Cabot Lodge. The third group comprised senators who did not support joining the League of Nations under any circumstance. This group, known as the Irreconcilables, was led by Senators William Borah, Robert La Follette, and Hiram Johnson.

Among the critics of the League of Nations, Senator Henry Cabot Lodge was the most vocal and influential (Cooper, 2010). Lodge was an eloquent speaker, prolific writer, and one of the first recipients of a doctorate in history from Harvard (Lodge, 1913; Biographical Directory of the United States

Congress, n.d.; The Senate Committee on Foreign Relations, 2000). From 1887 to 1893, Lodge was a member of the House of Representatives, and from 1893 to 1924, he was a senator. Moreover, he was appointed a member of the Senate Committee on Foreign Relations in 1895 and was its chairman from 1919 until his death on November 9, 1924.

According to Lodge, the opponents' main objection to the League of Nations was to Article X of the Covenant and the idea of maintaining international peace and security through collective actions (Lodge, 1925, p. 179). Opponents were also worried that joining the League of Nations would weaken the Monroe Doctrine. Further, some League opponents, including Lodge, were worried that joining the League of Nations would weaken Congress's legislative prerogatives. Despite all efforts, the Senate could not ratify the treaty (Cooper, 2010). As a result, the United States did not join the organization which it had itself proposed through its president.

Many scholars have blamed President Wilson for this failure. According to Cooper, "the League fight would almost certainly have resolved itself into some such compromise but for the actions of a single person. As everyone then and later recognized, that person was Woodrow Wilson" (Cooper, 2010, p. 2). Those who blame Wilson mainly point to his emotional disposition, his physical illness, and his psychiatric condition (Walker, 1995, p. 698). I, however, argue that failure of the League debate was because of fundamental differences between the strategic cultures of supporters and critics of the League.

The League debate was not the first time that Lodge and likeminded people had opposed US international commitments. In July 1911, Lodge and his friend President Theodore Roosevelt opposed proposals by President Henry Taft for US involvement in two General Arbitration Treaties with France and Great Britain (Hewes, 1970, p. 246). As with the case of the League of Nations, Lodge and Roosevelt opposed an open-ended entanglement with the outside world. Eventually, President Taft agreed to include Lodge's amendments to the arbitration treaties. However, Lodge's amendments and later modifications made by the Democrats in the Senate changed the treaties so much that Taft decided not to exchange their ratifications.

The shared factor in the 1911 and 1919 debates was different interpretations about the appropriate strategy of the United States (Self) in dealing with the world (Other). As Hewes argues, "[i]n the final analysis the conflict between the League's supporters and its critics involved a more fundamental one over the nature of force in international affairs" (Hewes, 1970, p. 249). The main goal of this chapter is to show there was a fundamental disagreement between supporters and opponents of the League of Nations regarding the utility of use of force (i.e. strategic culture), and that disagreement can explain the fate of the League debate in the United States.

Strategic Culture and its Microfoundations

Explaining and understanding relations between Self (the United States) and Other (the Soviet Union) and their similarities and differences was at the core of scholarly works on international relations during the Cold War. Some scholars, such as Kenneth Waltz (2010) and Thomas Schelling (1997), argued that the United States and the Soviet Union were not very different and that they were rational actors reacting to each other in an anarchic international system. Other scholars, such as Jack Snyder (1977) and Colin S. Gray (1981), argued that the United States was different from the Soviet Union and that difference was attributable to what they called strategic culture.

Snyder (1977) defined strategic culture as "the sum total of ideas, conditioned emotional responses, and patterns of habitual behavior that members of a national strategic community have acquired through instruction or imitation and share with each other" (p. 8). Similarly, Gray (1981) defined strategic culture as "modes of thought and action with respect to force" that are derived from "perception of the national historical experience" and "aspiration for self-characterization" (p. 22).

While the concept of strategic culture soon became popular both in academia and in policy, clarifying and operationalizing it proved difficult. To overcome methodological shortcomings, Johnston (1995) suggested two content analysis methods: cognitive mapping (Axelrod, 1976) and symbol analysis (Elder & Cobb, 1983). However, following Walker and Schafer (2007), I argue that the operational code method is a robust alternative for extracting the microfoundations of strategic culture. Johnston (1995) suggests that strategic culture should have two sets of assumptions: those about the "orderliness of the strategic environment" and those about the most efficacious means to operate in a strategic environment (p. 46). Assumptions about the orderliness of the strategic environment clarify the nature of the adversary and the usefulness of application of force (Johnston, 1995, pp. 44–55). Assumptions about efficacious means, on the other hand, should provide ranked preferences between conflictual strategies and cooperative/accommodationist strategies.

Operational code analysis has explicit indices for Johnston's suggested assumptions. Its philosophical indices extract the nature of adversary and its instrumental indices measure the usefulness of application of force. As will be explained in Section 4 of this chapter entitled "Subjective Games of Wilson and Lodge," using Theory of Inferences about Preferences (TIP), the operational code analysis also extracts actors' ranked preferences for strategic options. Another advantage of the operational code analysis is its ability to distinguish between strategic culture of the Self and strategic culture of the Other in a text. For instance, it can extract Soviet perception of its own strategic culture (Self) and the Soviet perception of the strategic culture of others (Other) from the texts published by Politburo. Further, Johnston also argues that it is important that the "content analysis of strategic cultural objects begins at the earliest point in history that is accessible to the researcher" (Johnston, 1995, p. 49). He also suggests that studies in strategic culture should be conducted both within a particular state and across states (Johnston, 1995, p. 54). While studying strategic culture on these levels and requirements seems daunting, and requires considerable resources, the automated operational code method, using the Verbs in Context System, makes achieving these goals attainable for scholars with limited resources.

Operational Code Method

The classic operational code method was introduced by Nathan Leites (1951; 1953) when he studied the Soviet elite. Later, Alexander George (1979) gave the operational code method more structure by introducing two sets of questions. Answers to the first set, the philosophical questions, reveal an actor's view about the strategic environment. The second set, the instrumental questions, present how an actor prefers to proceed in the strategic environment (George, 1979, pp. 201–216):

Alexander George's Philosophical Questions:

P1: What is the "essential" nature of political life? Is the political universe essentially one of harmony or conflict? What is the fundamental character of one's political opponents?

P2: What are the prospects for the eventual realization of one's fundamental political values and aspirations? Can one be optimistic, or must one be pessimistic on this score; and in what respects the one and/or the other?

P3: Is the political future predictable? In what sense and to what extent?

P4: How much "control" or "mastery" can one have over historical development? What is one's role in "moving" and "shaping" history in the desired direction?

P5: What is the role of "chance" in human affairs and in historical development?

Alexander George's Instrumental Questions:

I1: What is the best approach for selecting goals or objectives for political action?

I2: How are the goals of action pursued most effectively?

I3: How are the risks of political action calculated, controlled, and accepted?

I4: What is the best "timing" of action to advance one's interest?

I5 : What is the utility and role of different means for advancing one's interest?

In the 1990s, Walker, Schafer, and Young (1998) introduced the Verbs in Context System (VICS) for automatic extraction of answers to Alexander George's philosophical and instrumental questions from texts. VICS extracts the transitive verbs for Self and Other in a text and puts them into six categories of reward, promise, appeal/support, oppose/resist, threaten, and punish. In the next step, it assigns numeric values between (+3) to (-3) to each verb based on its intensity. In the end, using balance, central tendency, and variation of positive and negative verbs, it extracts answers to Alexander George's philosophical and instrumental questions. VICS extracts answers to philosophical questions by considering transitive verbs used for Other and it extracts answers to instrumental questions by considering transitive verbs used for Self.

VICS indices are also used to locate an actor in Holsti's belief system typologies. Holsti (1977) had introduced six typologies (A, B, C, D, E, F) for political leadership from which Walker and Schafer (2007, pp. 750–754) created a revised version by putting the pessimistic typologies (D, E, F) in one group (DEF) and, therefore, reducing the number of typologies to four (A, B, C, DEF). Type A and Type C both see the nature of politics and direction of strategy as cooperative. However, they differ in the amount of control that agents can exert over historical developments. Relative to Type A, Type C believes one has more control over historical developments. Type B and Type DEF see the nature of politics/direction of strategy as conflictual; however, they differ in control over the historical developments. Relative to Type DEF, Type B believes one has more control over historical developments.

Strategic Cultures of Wilson and Lodge

To compare Wilson's and Lodge's strategic cultures, their speeches and remarks in the period from 1913 to 1919 were gathered (spanning a year before and after WWI). Data for extracting Senator Lodge's strategic culture was compiled into three groups: (a) Lodge's speeches from 1913 to 1919, (b) his remarks about the League of Nations, and (c) his Senate floor remarks in 1919. Data for President Wilson was compiled into two groups: (a) his remarks from 1913 to 1919 and (b) a selection of his remarks about the League of Nations.

Texts were then analyzed using Profiler Plus, Version 7.3.11 (Social Science Automation, 2017), which automates the Verbs in Context System (VICS); only documents that had created at least six iterations of verbs for the Self were considered for analysis. President Wilson's remarks could generate enough texts that met the six-verb condition for all 7 years from 1913 to 1919. However, Senator Lodge's texts could only produce data for 1913, 1915, 1916, 1917, and 1919. To compare President Wilson's results with Senator Lodge's results, in the calculation of President Wilson's overall operational code, those years in which Senator Lodge had data were considered. The Appendix contains the raw results of the above analysis while Figure 6.1 and Figure 6.2, below, map the deviation of Wilson's and Lodge's instrumental and philosophical indices from the averages for a norming group of 255 speech acts by world leaders from different regions and historical eras (leaders' average). In the next section, the strategic cultures of Lodge and Wilson will be studied in two segments. The first segment will study Wilson's and Lodge's perception of strategic culture for Self by comparing their instrumental indices, and the second will compare Wilson's and Lodge's perception of strategic culture for Other using their philosophical indices.

Wilson's and Lodge's Strategic Cultures for Self

As Figure 6.1 shows, for instrumental indices (I indices), which measure microfoundations of strategic culture for Self, Lodge and Wilson show major differences from each other or the leaders' average in the following indices: strategic approach to goals (I1), risk orientation (I3), flexibility of tactics (I4a and I4b), utility of appealing (I5ap), utility of promising (I5pr), utility of opposing (I5op), utility of rewarding (I5re), and utility of threatening (I5th). The differences in Figure 6.1 are measured in standard deviations from the average scores for the norming group.

Relative to Wilson and the leaders' average, Lodge's strategic approach to achieving goals (I1) was less cooperative. Particularly, his direction of strategy was less cooperative when he referred to the League of Nations. On the other hand, Wilson's strategic approach to achieving goals was not only more cooperative than Lodge's, it was much more cooperative than the leaders' average as well. This shows that relative to Lodge, Wilson was much more eager to cooperate on the issue of the League of Nations.

Regarding risk orientation (I3), and relative to the leaders' average, Lodge was significantly more risk-averse both in his overall beliefs from 1913 to 1919 and when he spoke about the League of Nations. On the other hand, while Wilson's overall risk orientation from 1913 to 1919 was more risk-averse than the leaders' average, he was significantly more risk-acceptant when he spoke about the League of Nations. As Figure 6.1 shows, Wilson almost makes a complete turn from being risk-averse to being risk-acceptant. Risk-acceptance points to the acceptance of consequences associated with strategic choices (Walker, et al., 1998, p. 180). Lodge's relative risk-aversion shows he was not accepting the consequences of his relatively conflictual strategy. Wilson's high risk-acceptance about the League of Nations, however, shows he was ready to accept the risks of his relatively cooperative strategy.

For flexibility of tactics, VICS provides two measurements: I4a and I4b (Walker et al., 1998, p. 181). One index measures the propensity to shift between cooperative and conflictual tactics (I4a) and one measures the propensity to shift between words and deeds (I4b).

As Figure 6.1 shows, Lodge's propensity to shift between cooperative and conflictual tactics was much higher than the leaders' average, especially when he referred to the League of Nations. This shows his relatively conflictual strategy was not fixed and he was open to considering tactical cooperation. Regarding the propensity to shift between words and deeds (I4b) and relative to the leaders' average, Lodge shows a relatively high degree of flexibility, especially when referring to the League of Nations. This high level of propensity to shift between words and deeds shows he did not accept the risk associated with his choice of means (threat, punish, oppose vs. reward, appeal, promise), and, therefore, he was risk-averse when the outcome was submission or when it was a deadlock. The combination of these two indices indicates Lodge was inclined to consider cooperation but did not want to be completely cooperative. At the same time, he did not want to be completely uncooperative either. As a Reservationist, he was inclined to accept some tactical cooperation.

Figure 6.1

Instrumental Beliefs of President Wilson and Senator Lodge

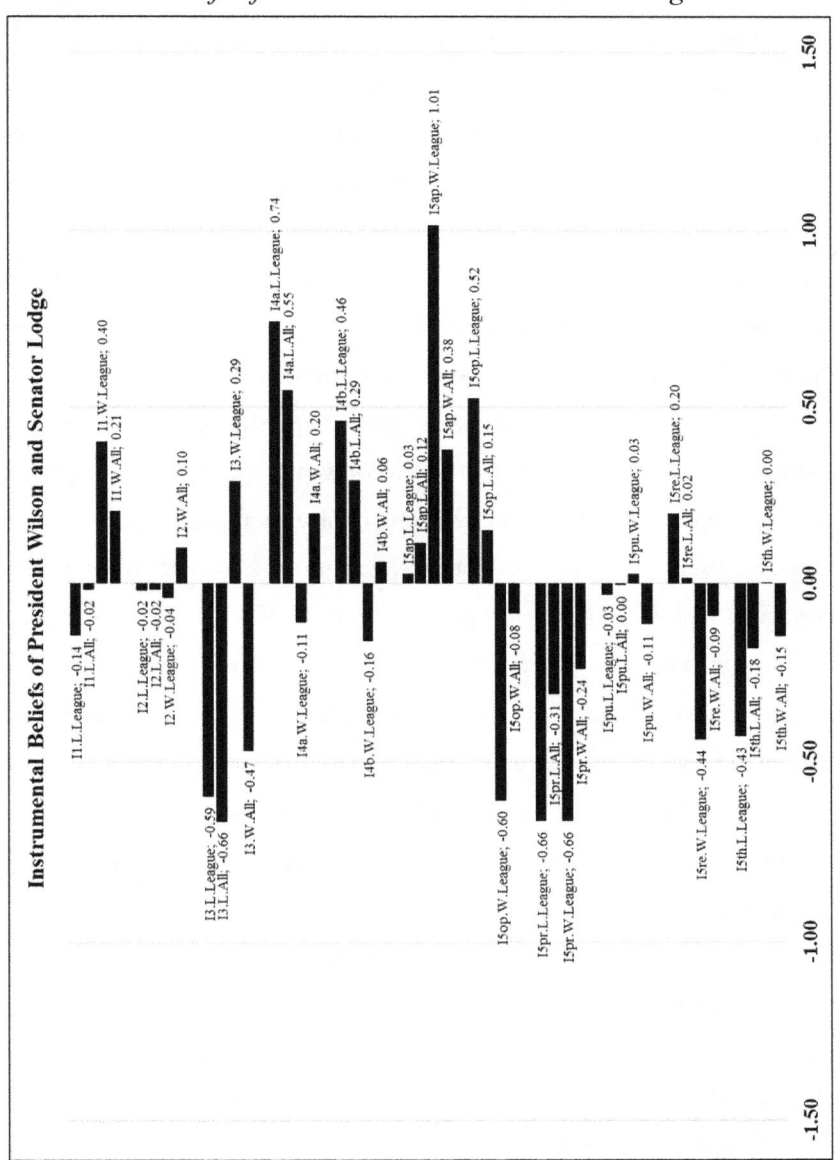

Note. Indices are shown in standard deviations from the average scores for the norming group of 255 speech acts by world leaders from different regions and historical eras. All = Data for 1913, 1915, 1916, 1917, and 1919; League = Data for remarks about League of Nations; L = Lodge; W = Wilson; ap = appeal; op = oppose; pr = promise; pu = punish; re=reward; th = threaten.

On the other hand, when Wilson referred to the League of Nations, relative to Lodge, to the leaders' average, and even to his overall view, he showed less inclination to shift between conflict and cooperation (I4a). This shows that his relatively strong cooperative strategy was real. Regarding the propensity to shift between words and deeds (I4b), Wilson was less flexible than Lodge, showing that he was better prepared to accept the risks associated with his choice of means (threat, punish, oppose vs. reward, appeal, promise). Speaking about the League of Nations made Wilson even more resolute on cooperation. This index shows that for the League of Nations, Wilson was willing to be very cooperative.

Regarding the choice of means (threat, punish, oppose vs. appeal, promise, reward) and relative to the leaders' average, Lodge preferred to use fewer appeals than Wilson. Wilson's choice to appeal, especially when speaking about the League of Nations, was much higher than the leaders' average as well. In choosing to oppose as a means of achieving goals, Wilson and Lodge were completely different. Lodge opposed more frequently, particularly when referring to the League of Nations. They were also different in their choice of reward and threat as appropriate means of achieving goals. Speaking about the League of Nations, Lodge tended to reward more and threaten less. Wilson, on the other hand, preferred to reward less and threaten more.

The above findings show that, concerning the Self component of strategic culture, Wilson and Lodge differed greatly. This difference can be seen both over time (1913–1919) and when they talked about the League of Nations. This finding strongly supports Hypothesis 1. Wilson and Lodge's strategic cultures for Self were different. The above findings also support Hypothesis 2. Lodge's strategic culture for Self was relatively more hostile and less cooperative than Wilson's. However, Lodge also showed some inclinations for tactical cooperation.

Wilson's and Lodge's Strategic Cultures of Other

Figure 6.2 presents the philosophical indices (P indices), which measure microfoundations of strategic culture for the Other. As Figure 6.2 indicates, Lodge and Wilson somewhat deviated from each other or from the leaders' average in all the philosophical indices: nature of the political universe (P1), realization of political values (P2), predictability of political future (P3), control over historical development (P4a and P4b), and role of chance (P5).

Concerning the nature of politics (P1) and relative to the leaders' average, Lodge's overall view (1913–1919) was less friendly. However, when he referred to the League of Nations, he presented a very friendly view of the world. Wilson, on the other hand, consistently showed a friendlier view of the world relative to the leaders' average and Lodge. In general and relative to the leaders' average, Wilson and Lodge were more pessimistic in realizing their political values (P2). However, when referring to the League of Nations, they both became more optimistic than the leaders' average, with Wilson showing a higher degree of optimism. For the predictability of the political future (P3) and relative to the leaders' average, the overall view of both Wilson and Lodge was that the future was not that predictable, with Lodge having an uncertain view of the future. However, talking about the League of Nations, they both became more optimistic than the leaders' average, with Wilson seeing the future as being significantly more predictable than did Lodge. VICS has two indices for control over historical development. One (P4a or P4Self) measures control of Self over historical development, and the second (P4b or P4Other) measures Other's control. As Figure 6.2 shows, Wilson and Lodge assign less control to Self than the leaders' average. On the other hand, relative to the leaders' average, they both assign more control to Other (P4Other). Similarly, relative to the leaders' average, and across the board, Wilson and Lodge assign more to the role of chance (P5).

Figure 6.2

Philosophical Beliefs for President Wilson and Senator Lodge

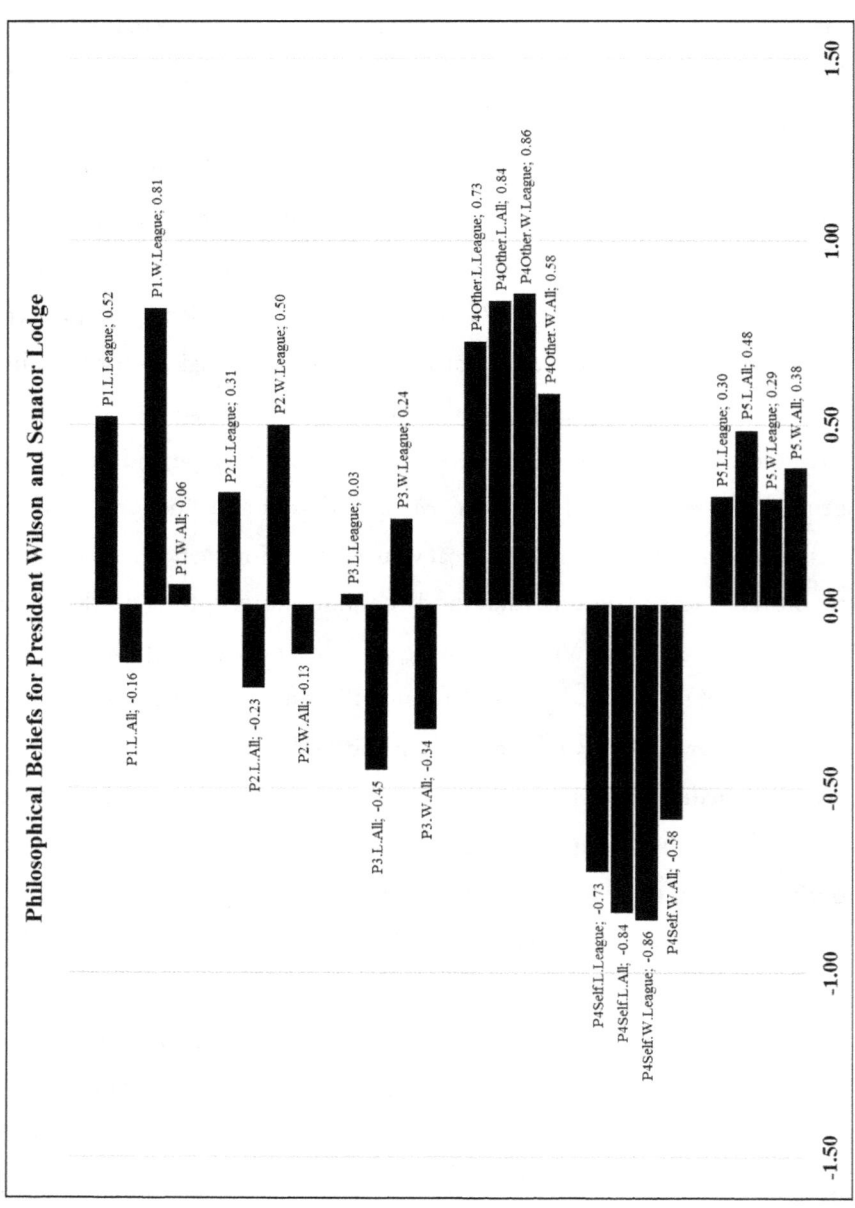

Note. Indices are shown in standard deviations from the average scores for the norming group of 255 speech acts by world leaders from different regions and historical eras. All = Data for 1913, 1915, 1916, 1917, and 1919; League = Data for remarks about League of Nations; L = Lodge; W = Wilson

The findings above show that, for the Other component of strategic culture, Wilson and Lodge were different but not as much with the Self component of strategic culture. This finding somewhat supports Hypothesis 1. They had more differences when compared to the leaders' average than with each other. However, the above findings support Hypothesis 2. Lodge's strategic culture for Other was relatively less friendly and more pessimistic than Wilson's.

Subjective Games of Wilson and Lodge

In the previous section, I evaluated Hypothesis 1 and Hypothesis 2 by extracting and comparing the strategic culture of Wilson and Lodge for Self and Other. The question is whether their different strategic cultures can explain the result of the debate about the League of Nations, which was a deadlock. The third hypothesis gives an affirmative answer to this question. To evaluate Hypothesis 3, I first extract Wilson's and Lodge's ranking of preferred outcomes. Then, I present the subjective games of Wilson and Lodge and their dominant strategies.

Ranking of Preferred Outcomes

To extract and rank preferred political outcomes, the operational code method uses the indices for nature of the political universe (P1), strategic orientation (I1), Self's historical control (P4a or P4Self), and Other's historical control (P4b or P4Other). The (I1, P4Self) pair maps the ranking order of the preferred outcomes for Self. The (P1, P4Other) pair maps the ranking order of the preferred outcomes for Other.

To present Lodge's and Wilson's choice preferences across time, I map their preferences in a 2x2 matrix using Holsti's revised belief system typology, with I1 and P1 in the vertical axis and P4a and P4b in the horizontal axis. To better map agents' preferences, instead of their raw indices, I use the deviation of the raw indices from the leaders' average (z-score). The intersections of vertical and horizontal indices "specify what sequence of behaviors is appropriate for Self to pursue and what sequence of behaviors can be expected from Other" (Walker & Schafer, 2018).

To extract Wilson and Lodge's ranked preferences, I use the Theory of Inferences about Preferences (TIP). Based on the deviation of I1, P1, P4a and P4b from the leaders' average, TIP reveals the ranking of preferred outcomes for Wilson and Lodge:

> When the [I1 or P1] index scores lie above this [leaders'] norm, the index is considered "+" for the purposes of inferring preferences. Where they lie below the norm they are considered "-" for that purpose. For the P-4 index, a plus or minus one standard deviation norming range was set. If an actor's P-4 VICS score falls within this range, it is considered "="; otherwise, it is either above ">" or below "<." (Marfleet & Miller, 2005, pp. 344–345)

Based on the above rules, TIP offers the following propositions which reveal the ranking order of preferences and strategies associated with them (Malici, 2011, p. 91):

1. *Appeasement Strategy*: If (I-1, P-4a) or (P-1, P-4b) is (+, <), then Settle > Deadlock > Submit > Dominate.
2. *Assurance Strategy*: If (I-1, P-4a) or (P-1, P-4b) is (+, =), then Settle > Deadlock > Dominate > Submit.
3. *Stag Hunt Strategy*: If (I-1, P-4a) or (P-1, P-4b) is (+, >), then Settle > Dominate > Deadlock > Submit.
4. *Chicken Strategy*: If (I-1, P-4a) or (P-1, P-4b) is (–, <), then

Dominate > Settle > Submit > Deadlock.

5. *Prisoner's Dilemma Strategy*: If (I-1, P-4a) or (P-1, P-4b) is (–, =), then Dominate > Settle > Deadlock > Submit.

6. *Bully Strategy*: If (I-1, P-4a) or (P-1, P-4b) is (–, >), then Dominate > Deadlock > Settle > Submit.

Lodge: Belief System Typology and Preferred Political Outcomes

Figures 3 and 4 map Self and Other images for Lodge (L) and Wilson (W) over time. Figure 6.3 shows that Lodge's view of Self is a Type DEF both in his overall view from 1913 to 1919 and when he talks about the League of Nations. From the eight instances of measuring Lodge's view, only two (L.1917.S and L.Senate.S) are not in the DEF quadrant, and one of the two (L.Senate.S) is extremely close to the DEF quadrant. This shows his expressed view of Self, when referring to the League of Nations, was in accordance with his overall view across time. Figure 6.3 also presents Theodore Roosevelt's views about the League of Nations, which were extracted from his remarks about this topic. As Figure 6.3 shows, Lodge's view of Self is also similar to the view of his like-minded friend, Theodore Roosevelt. This expected similarity between Lodge and Roosevelt increases the confidence in measuring Lodge's strategic culture for Self. Based on Lodge's values for I1 and P4Self in Figure 6.3 and according to the above TIP propositions, Lodge's strategy for Self in dealing with the issue of the League of Nations was the prisoner's dilemma strategy. Accordingly, his ranking order for Self's preferences would be Dominate > Settle > Deadlock > Submit. As most of Lodge's I1 and P4Self values meet the same TIP conditions, there is a high degree of confidence that Lodge's prisoner's dilemma strategy for Self in dealing with the issue of the League of Nations was his dominant strategy for Self across time as well.

Regarding Lodge's images of Other, as Figure 6.3 shows, his view was not consistent over time. His overall view of Other from 1913 to 1919 puts

Other in the Type B quadrant. One expects that, as Lodge and Theodore Roosevelt presented a similar view of Self, their view of Other would be similar as well. However, as Figure 6.3 shows, concerning the League of Nations, Lodge puts Other in the Type C quadrant and Roosevelt puts Other in the Type B quadrant. Talking about the League of Nations, Roosevelt is more similar to Lodge's overall view of Other than Lodge himself. Based on Lodge's values of P1 and P4Other as shown in Figure 6.3 and according to the above TIP propositions, Lodge's strategy for Other concerning the League of Nations was the assurance strategy. Accordingly, his ranking order for Other's preferences would be Settle > Deadlock > Dominate > Submit.

Wilson: Belief System Typology and Preferred Political Outcomes

Wilson, on the other hand, locates Self in the Type A quadrant across time. As Figure 6.4 shows, from nine instances of measuring Wilson's view of Self, all are located in the Type A quadrants, which indicates consistency in Wilson's view of Self and increases confidence in the measurement. Based on the values of I1 and P4Self in Figure 6.4 and according to the above TIP propositions, Wilson's strategy for Self concerning the League of Nations was the assurance strategy. Accordingly, his ranking order for Self's preferences would be Settle > Deadlock > Dominate > Submit. As all of Wilson's I1 and P4Self values meet the same TIP conditions, there is a high degree of confidence that Wilson's assurance strategy for Self in dealing with the issue of the League of Nations was his dominant strategy for Self across time as well.

Wilson's overall view of Other shows much more consistency than Lodge's view (Figure 6.4). From the nine instances of measuring Wilson, eight are located in the Type C quadrant, including his overall view and his view of Other when he refers to the League of Nations. Based on the values of P1 and P4Other in Figure 6.4 and according to the above TIP

Figure 6.3

Self and Other Images for Senator Lodge

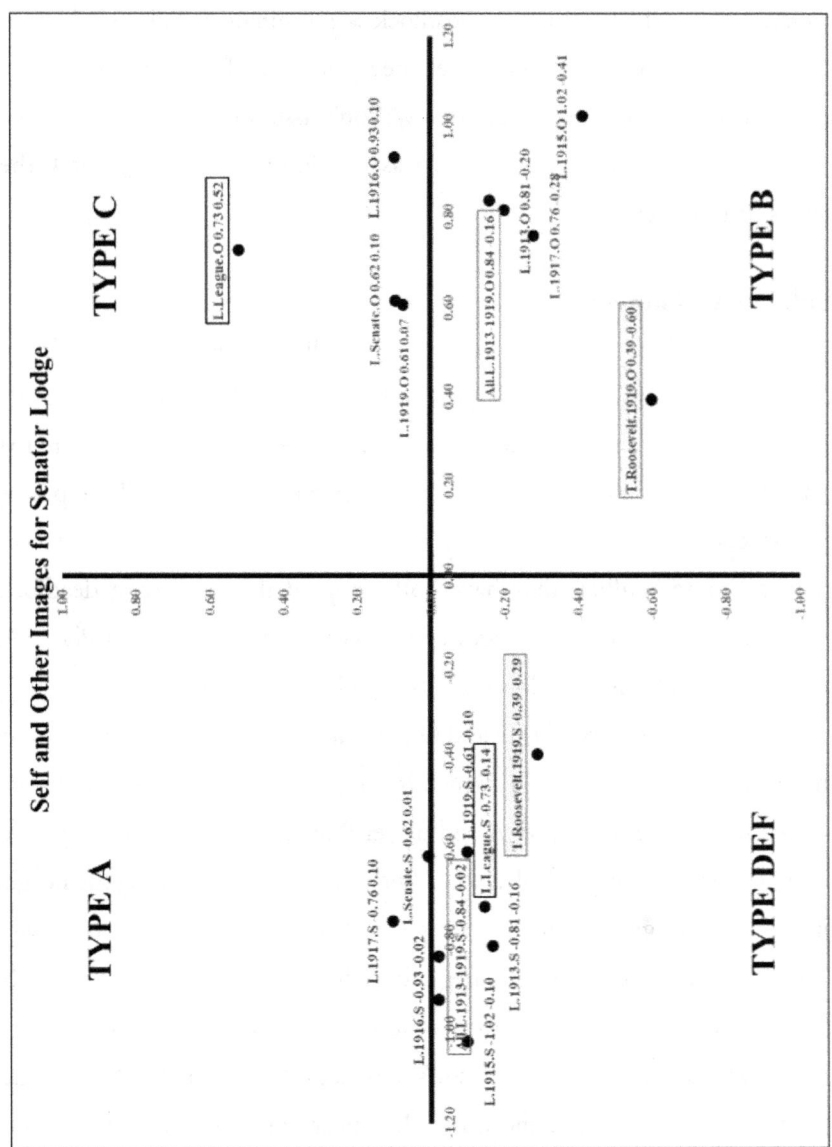

Note. Indices are shown in standard deviations from the average scores for the norming group of 255 speech acts by world leaders from different regions and historical eras. I1 and P1 are in the vertical axis and P4a and P4b are in the horizontal axis. All = Lodge's Data for 1913, 1915, 1916, 1917, and 1919; League = Data for remarks about League of Nations; L = Lodge; O = Other; S = Self propositions, Wilson's strategy for Other concerning the League of Nations was the assurance strategy. Accordingly, his ranking order for Other's preferences would be Settle > Deadlock > Dominate > Submit. As most of Wilson's P1 and P4Other values meet the same TIP conditions, there is a high degree of confidence that Wilson's assurance strategy for Other concerning the League of Nations was his dominant strategy for Other across time as well.

Subjective Games

Using Lodge's and Wilson's preferred political outcomes for Self and Other, this section presents Wilson and Lodge's duel over the issue of the League of Nations in 2x2 ordinal games. In these 2x2 games, there are two players and each player has two options: cooperate or conflict. If both players choose cooperate, they have both adopted the strategy of settle. If both players choose conflict, they have both adopted the strategy of deadlock. If Player A chooses cooperate and Player B chooses conflict, then Player A's strategy is to submit and Player B's strategy is to dominate.

Figure 6.5 presents Wilson and Lodge's duels over the issue of the League in two games. In the Game 1, which is Lodge's game in the League debate, ranked preferences for player A and player B are Lodge's ranked preferences for Self and Other respectively. According to Game 1, concerning the League of Nations, Lodge's dominant strategy for Self is conflict. No matter what Other chooses, Lodge's best strategy for Self is to be uncooperative. In the Game 2, which is Wilson's game in the League debate, ranked preferences for Player A and Player B are Wilson's ranked preferences for Self and Other respectively. According to the Game 2, concerning the League of Nations, Wilson does not have a dominant strategy and his strategy depends on the strategy of Other. If Other chooses to cooperate, Wilson's best strategy for Self is to cooperate. If Other chooses conflict, then Wilson's best strategy for

Figure 6.4

Self and Other Images for President Wilson

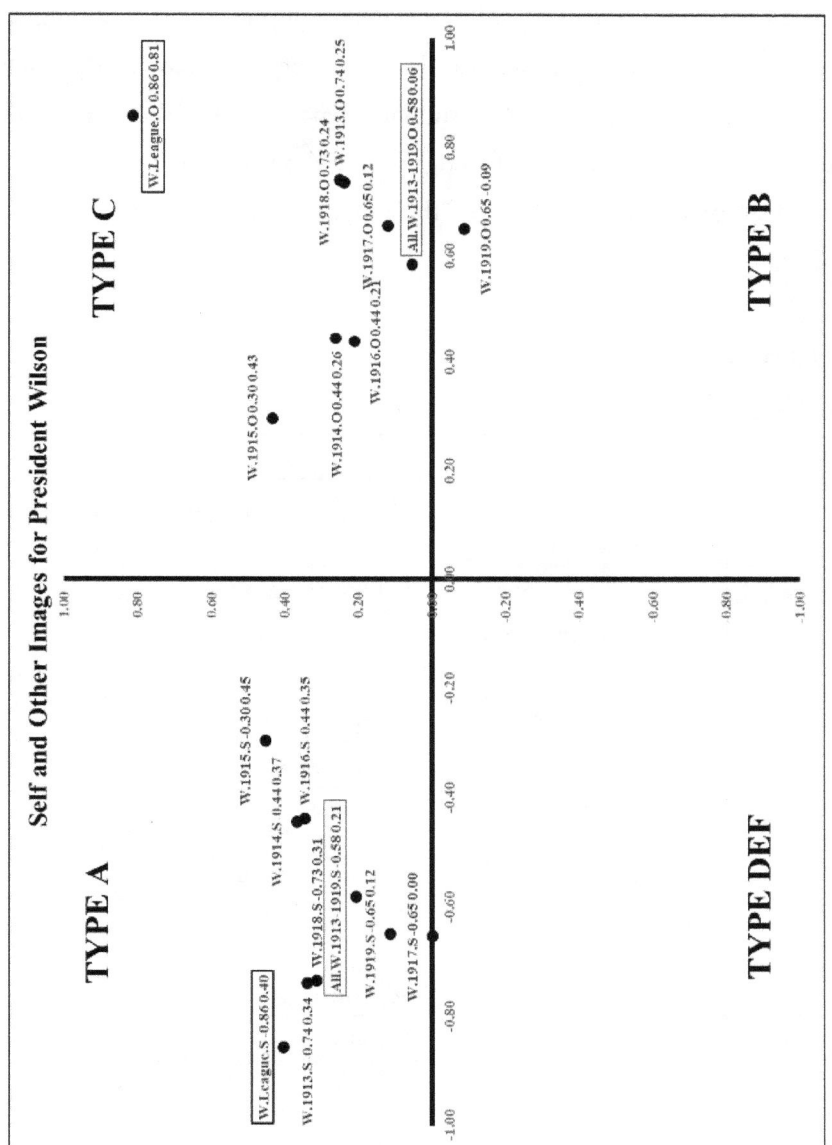

Note. Indices are shown in standard deviations from the average scores for the norming group of 255 speech acts by world leaders from different regions and historical eras. I1 and P1 are in the vertical axis and P4a and P4b are in the horizontal axis. All = Wilson's Data for 1913, 1915, 1916, 1917, and 1919; League = Date for remarks about League of Nations; O = Other; S = Self; W = Wilson.

Self is conflict. As Lodge's dominant strategy is conflict, Wilson's contingent strategy has to be conflict as well. The above two games show that owing to the ranking of preferred political outcomes by Lodge and Wilson, the deadlock in the League debate was the most expected outcome. The results of the above two games strongly support Hypothesis 3 of this chapter: the extracted strategic cultures of Lodge and Wilson for Self and Other can correctly predict the deadlock of the League debate between Lodge and Wilson.

Figure 6.5

Subjective Games of Senator Lodge and President Wilson

		Player B	
		Cooperate	Conflict
Player A	Cooperate	Settle	Submit
	Conflict	Dominate	Deadlock
Sample Game: Outcomes for Player A			

		Lodge/Other	
		Cooperate	Conflict
Lodge/Self	Cooperate	3,4	1,2
	Conflict	4,1	2,3[a]
Game 1: Senator Lodge			

		Wilson/Other	
		Cooperate	Conflict
Wilson/Self	Cooperate	4,4[a]	1,2
	Conflict	2,1	3,3[a]
Game 2: President Wilson			

Note. Ranked preferences for Player A and Player B are ranked preferences for Self and Other respectively.
[a]. Nash Equilibrium in Pure Strategies

Conclusion

In some ways, this chapter vindicated President Wilson and presented an alternative picture of the League debate—a picture in which Wilson is not an uncompromising character but a leader who is open to cooperation. Wilson took many cooperative steps. For instance, to accommodate Republicans in the Congress, he successfully convinced other great powers to explicitly recognize the Monroe Doctrine in the text of the Covenant:

> What are those who advise us to turn away from it afraid of? In the first place, they are afraid that it impairs in some way that long traditional policy of the United States which was embodied in the Monroe doctrine, but how they can fear that I can not conceive, for the document expressly says in words which I am now quoting that nothing in this covenant shall be held to affect the validity of the Monroe doctrine. The phrase was inserted under my own eye, at the suggestion - not of the phrase but the principle - of the Foreign Relations Committees of both Houses of Congress. I think I am justified in dismissing all fear that the Monroe doctrine is in the least impaired. (Wilson, 1919)

Even Lodge himself confesses that Wilson cooperated a lot, as he told the Speaker of the House, Frederick Gillett, on July 26, 1920:

> They [the Democrats] insisted upon some form of reservation which would leave an obligation in existence. We were determined that there should be no obligation of any sort left under Article X. They went pretty far in offering exceptions to the obligation but they kept the obligation alive. (Link, 1991, as cited in Cooper, 2010, p. 308)

Lodge gambled on his perception of Wilson. He thought that Wilson would submit in the face of Lodge's dominant strategy of conflict. However, he was wrong:

> If President Wilson had been a true idealist, in regard to the covenant of the League of Nations, for example, he would have saved his covenant and secured its adoption by the Senate of the United States by accepting some modification of its terms, since the man who really seeks the establishment of an ideal will never sacrifice it because he cannot secure everything he wants at once, and always estimates the principle as more important than its details and qualifications. (Lodge, 1925, p. 225)

References

Axelrod, R. M. (1976). *Structure of Decision: The Cognitive Maps of Political Elites*. Princeton: Princeton University Press.

Biographical Directory of the United States Congress (n.d.) *LODGE, Henry Cabot, (1850 – 1924)*. Retrieved from http://bioguide.congress.gov/scripts/biodisplay.pl?index=L000393

Cooper, J. M. (2010*). Breaking the Heart of the World: Woodrow Wilson and the Fight for the League of Nations.* Cambridge: Cambridge University Press.

Elder, C. D., & Cobb, R. W. (1983). *The Political Use of Symbols*. New York: Longman.

George, A. L. (June 01, 1969). The "Operational Code": A Neglected Approach to the Study of Political Leaders and Decision-Making. *International Studies Quarterly*, 13, 2, 190-222.

Gray, C. S. (1981). National Style in Strategy: The American Example. *International Security*, 6(2), 21-47.

Hewes, J. E. (August 20, 1970). Henry Cabot Lodge and the League of Nations. *Proceedings of the American Philosophical Society*, 114(4), 245-255.

Holsti, O. (1977). *The Operational Code as an Approach to the Analysis of Belief Systems: Final Report to the National Science Foundation Grant No. SOC75*-15368. Durham: Duke University Press.

Johnston, A. I. (January 01, 1995). Thinking About Strategic Culture. *International Security, 19*(4), 32-64.

Leites, N. (1951). *The Operational Code of the Politburo*. New York: McGraw-Hill.

Leites, N. (1953). *A Study of Bolshevism*. Glencoe, Illinois: Free Press.

Lodge, H. C. (1913). *Early Memories*. New York: Charles Scribner's Sons.

Lodge, H. C. (1925). *The Senate and the League of Nations*. New York: Charles Scribner's Sons.

Malici, A. (2011). The United States and Rogue Leaders: Understanding Conflicts. In S. G. Walker, A. Malici & M. Schafer. *Rethinking Foreign Policy Analysis: States, Leaders, and the Microfoundations of Behavioral International Relations*. (pp. 83-96). London: Routledge.

Marfleet, G., & Miller, C. (November 01, 2005) Failure after 1441: Bush and Chirac in the UN Security Council. *Foreign Policy Analysis* 1, 333–360.

Schelling, T. C. (1997). *Strategy of Conflict*. London: Harvard University.

The Senate Committee on Foreign Relations. (October 2000). *Millennium Edition 1816–2000*. Retrieved from https://www.foreign.senate.gov/imo/media/doc/CDOC-105sdoc281.pdf

Snyder, J. L. (1977). *The Soviet Strategic Culture: Implications for Limited Nuclear Operations*. Retived from https://www.rand.org/content/dam/rand/pubs/reports/2005/R2154.pdf

Social Science Automation. (2017). Profiler Plus [Computer Software]. Retrieved From https://profilerplus.org/default.aspx

The United States Senate (n.d.). *Woodrow Wilson Addresses the Senate*. Retrieved from https://www.senate.gov/about/powers-procedures/treaties/woodrow-wilson-addresses-the-senate.htm.

The Versailles Treaty (June 28, 1919). Retrieved from https://avalon.law.yale.edu/subject_menus/versailles_menu.asp.

Walker, S. (December 01, 1995). Psychodynamic Processes and Framing Effects in Foreign Policy Decision-Making: Woodrow Wilson's Operational Code. *Political Psychology*, 16, 4, 697-717.

Walker, S., & Schafer, M. (December 01, 2007). Theodore Roosevelt and Woodrow Wilson as Cultural Icons of U.S. Foreign Policy. *Political Psychology*, 28, 6, 747-776.

Walker, S., & Schafer, M. (January 11, 2018). *Operational Code Theory: Beliefs and Foreign Policy Decisions*. Retrieved from http://internationalstudies.oxfordre.com/view/10.1093/acrefore/9780190846626.001.0001/acrefore-9780190846626-e-411.

Walker, S., Schafer, M., & Young, M. (1998). Systematic Procedures for Operational Code Analysis. *International Studies Quarterly*, 42, 175-189.

Waltz, K. N. (2010). *Theory of International Politics*. Long Grove, Ill: Waveland Press.

Wilson, W. (January 8, 1918). *President Woodrow Wilson's Fourteen Points*. Retrieved from http://avalon.law.yale.edu/20th_century/wilson14.asp.

Wilson, W. (September 19, 1919). *Address at the Stadium in Balboa Park in San Diego, California*. Retrieved from http://www.presidency.ucsb.edu/ws/index.php?pid=117390.

World Peace Foundation. (1919). *League of Nations: Volume II*. Boston: World Peace Foundation.

Appendix
Operational Codes of Senator Lodge and President Wilson
Table A1
Senator Lodge's Operational Codes

		Leaders' Average (n=255)	1913	1915	1916	1917	1919	1913-1919	Senate Debate 1919	League of Nations
	Instrumental Beliefs									
I-1	Strategic Approach to Goals (Cooperative/Conflictual)	0.33	0.26	0.29	0.33	0.38	0.29	0.33	0.34	0.27
I-2	Tactical Pursuit of Goals (Cooperative/Conflictual)	0.14	0.10	0.11	0.18	0.12	0.15	0.13	0.14	0.13
I-3	Risk Orientation (Averse/Acceptant)	0.30	0.19	0.16	0.10	0.22	0.09	0.14	0.16	0.16
I-4	Timing of Action									
	a. Cooperation/Conflict	0.51	0.74	0.71	0.67	0.62	0.71	0.67	0.66	0.73
	b. Words/Deeds	0.53	0.46	0.63	0.74	0.52	0.69	0.61	0.64	0.67
I-5	Utility of Means									
	a. Appeal/Support	0.43	0.49	0.46	0.38	0.54	0.37	0.45	0.47	0.43
	b. Promise	0.07	0.03	0.03	0.04	0.02	0.08	0.04	0.03	0.00
	c. Reward	0.17	0.11	0.16	0.24	0.13	0.20	0.17	0.17	0.20
	d. Oppose/Resist	0.15	0.26	0.20	0.17	0.13	0.19	0.17	0.17	0.23
	e. Threaten	0.05	0.00	0.00	0.03	0.05	0.02	0.03	0.01	0.00
	f. Punish	0.14	0.11	0.16	0.13	0.13	0.14	0.14	0.15	0.13
	Philosophical Beliefs									
P-1.	Nature of the Political Universe (Friendly/Hostile)	0.25	0.19	0.12	0.28	0.16	0.27	0.20	0.28	0.42
P-2.	Realization of Political Values (Pessimistic/Optimistic)	0.12	0.07	0.01	0.14	0.01	0.12	0.06	0.10	0.19
P-3.	Predictability of Political Future (Low/High)	0.15	0.10	0.10	0.11	0.09	0.09	0.09	0.12	0.15
P-4	Control Over Historical Development (Low/High)									
	a. Self's Control	0.21	0.11	0.09	0.10	0.12	0.14	0.11	0.14	0.12
	b. Other's Control	0.79	0.89	0.91	0.90	0.88	0.86	0.89	0.86	0.88
P-5	Role of Chance	0.97	0.99	0.99	0.99	0.99	0.99	0.99	0.98	0.98

Note. Only documents that created at least six iterations of transitive verbs for the Self were included. Senator Lodge's operational code indices are based on his remarks in 1913, 1915, 1916, 1917, and 1919. 1913-1919 indices refer to Senator Lodge's overall operational code in 1913, 1915, 1916, 1917, and 1919. Senate Debate 1919 indices are from Senator Lodge's Senate floor remarks in 1919. League of Nations indices are based on Senator Lodge's remarks about the League of Nations. Leader's Average indices are from the averages for a norming group of 255 speech acts by world leaders from different regions and historical eras.

Table A2

President Wilson's Operational Codes

		Leaders' Average (n=255)	1913	1914	1915	1916	1917	1918	1919	1913-1919	League of Nations
	Instrumental Beliefs										
I-1	Strategic Approach to Goals (Cooperative/Conflictual)	0.33	0.49	0.51	0.55	0.50	0.33	0.48	0.39	0.44	0.52
I-2	Tactical Pursuit of Goals (Cooperative/Conflictual)	0.14	0.25	0.22	0.25	0.22	0.07	0.21	0.14	0.17	0.13
I-3	Risk Orientation (Averse/Acceptant)	0.30	0.17	0.28	0.25	0.23	0.21	0.22	0.17	0.20	0.37
I-4	Timing of Action										
	a. Cooperation/Conflict	0.51	0.51	0.49	0.45	0.50	0.67	0.52	0.61	0.56	0.48
	b. Words/Deeds	0.53	0.55	0.41	0.47	0.48	0.65	0.52	0.57	0.53	0.48
I-5	Utility of Means										
	a. Appeal/Support	0.43	0.49	0.59	0.57	0.56	0.52	0.54	0.49	0.53	0.67
	b. Promise	0.07	0.05	0.02	0.04	0.03	0.01	0.04	0.06	0.04	0.00
	c. Reward	0.17	0.20	0.14	0.17	0.16	0.13	0.17	0.14	0.15	0.10
	d. Oppose/Resist	0.15	0.13	0.17	0.12	0.13	0.12	0.14	0.14	0.14	0.05
	e. Threaten	0.05	0.04	0.01	0.04	0.04	0.02	0.03	0.03	0.03	0.05
	f. Punish	0.14	0.08	0.06	0.07	0.08	0.19	0.09	0.14	0.12	0.14
	Philosophical Beliefs										
P-1	Nature of the Political Universe (Friendly/Hostile)	0.25	0.33	0.33	0.39	0.32	0.29	0.33	0.22	0.27	0.51
P-2	Realization of Political Values (Pessimistic/Optimistic)	0.12	0.16	0.17	0.16	0.13	0.10	0.11	0.05	0.09	0.24
P-3	Predictability of Political Future (Low/High)	0.15	0.12	0.10	0.15	0.12	0.11	0.14	0.10	0.11	0.18
P-4	Control Over Historical Development (Low/High)										
	a. Self's Control	0.21	0.12	0.16	0.18	0.16	0.13	0.12	0.13	0.14	0.11
	b. Other's Control	0.79	0.88	0.84	0.82	0.84	0.87	0.88	0.87	0.86	0.89
P-5	Role of Chance	0.97	0.99	0.98	0.97	0.98	0.99	0.98	0.99	0.98	0.98

Note. Only documents that created at least six iterations of transitive verbs for the Self were included. President Wilson's operational code indices are based on his remarks in 1913, 1914, 1915, 1916, 1917, 1918, 1919. Because Senator Lodge's texts could only produce qualified data for 1913, 1915, 1916, 1917, and 1919, in the calculation of President Wilson's overall operational code (1913-1919), those years in which Senator Lodge had data were considered. League of Nations indices are from a selection of

President Wilson's remarks about the League of Nations. Leader's Average indices are from the averages for a norming group of 255 speech acts by world leaders from different regions and historical eras.

7

One Hundred Years On: The Shadow of League of Nations Failure on American Support for International Law

Jonathan S. Miner
University of North Georgia

Abstract

Why does the United States possess an uneven record of participation in international law and a modern reluctance to adopt treaties as the domestic law of the land? Given the importance of global cooperation in the rise of the United States to superpower status and its primary position as an original member of many important treaties, it is surprising that American support for international law has not been consistently higher over the past 100 years. This paper explores complementary literatures in international law, international relations, and the foreign policy making process and utilizes a latent content analysis of sources dating back to the 1919 failure of the League of Nations treaty to uncover possible explanations for this inconsistency. The research finds that in the United States, traditional support for international law exists alongside an equally influential history of wariness to enter such agreements and a practical skepticism for their usefulness, opposing forces that continually conflict within the constitutionally mandated treaty ratification process. Unlike automatic treaty ratification in many countries, it is within the separate, additional, and constitutionally mandated ratification process in the United States where this conflict persists. This research discovers a historical use of the League of Nations treaty failure as a rallying cry by opponents to defeat current legislation under consideration in the ratification process, building over time and resulting in the weakening of US support for international law.

Keywords: International law, League of Nations, United States foreign policy

One Hundred Years On: The Shadow of League of Nations Failure on American Support for International Law

The United States' participation in international law has a long and winding history. An important participant since the expansion of international law in the late 19th century, the United States has often been at the forefront of international cooperation. As illustrated by its status as a founding member of early treaties such as the Universal Postal Union of 1874 and Hague Conventions of 1899 and 1907, the United States was in the vanguard of international law. Twentieth century landmark treaties illustrating this leadership include the United Nations Charter (1945), the Bretton Woods international economic structure (1944), Universal Declaration on Human Rights (1948), and nuclear control and reduction treaties with the Soviet Union/Russian Federation since the 1970s.

Yet, the United States has also often displayed a disregard for participation in and adherence to international law, with notable opt-outs including the 1982 Convention on the Law of the Sea, Kyoto Protocol (1997), jurisdiction in the International Criminal Court (1998), withdrawal from the Anti-Ballistic Missile Treaty (2002), and the focus of this paper, the League of Nations Treaty (1919). This uneven participation in global governance seems surprising given the simultaneous rise of international law with the emerging dominance of the United States in the international system. Why would a global hegemon, a direct and substantial beneficiary of the benefits of international law and cooperation, also have such an uneasy and even conflicted relationship? Has US participation always come with this subtle, accompanying skepticism and disregard? Did the United States choose to act alone once it gained substantial benefit from international cooperation and emerged as a global power? Is it due to domestic political issues or popular opposition to international law at home? In sum, how does a student of international relations and law interpret almost 150 years of inconsistent US involvement in international law?

This paper seeks to investigate these questions, survey the relevant literature, and analyze evidence in this complicated relationship. The complementary literatures concerning the concepts surrounding state positions regarding adoption of international law, international relations theory, and US foreign policy making combine to suggest several interesting conclusions contributing to scholarship on the topic. First, this research concludes that in the United States, liberalist and Wilsonian support for international law exists alongside a separate undercurrent of distrust, isolationism, and skepticism over its practical outcomes. Second, the constitutional requirement of Senate ratification of any signed international treaty creates a contentious domestic political bargaining process, which adds significant risk to its potential to become the law of the land. Third, and perhaps most surprisingly, the failure to ratify and the failure of the League of Nations treaty continues to be the example most used as an illustration to oppose participation in international law.

Literature Review

The issues surrounding the contentious relationship between the United States and international law span several different literatures intersecting international law, international relations, and the foreign policy making process. To date, these literatures are just starting to combine, complement, and analyze these issues, and this paper intends to contribute to this growing area of study. It does so by exploring scholars' work in the following fields: (1) the international relations theory subfield of realism, (2) international legal realism, (3) domestic constitutional issues of treaty incorporation, and (4) the contentious, dual-level process of making foreign policy in the United States. While rarely studied in this combination, these questions become increasingly important as the world becomes more intertwined and the United States deeply reconsiders its place in that world.

The International Relations Subfield of Realism and International Law

Shirley V. Scott's (2004) analysis in "Is There Room for International Law in Realpolitik?—Accounting for the US Attitude towards International Law" is a good starting place in this literature review, and it was written a decade before scholars' work in the next section of this review began concerning legal realism in earnest. Scott asks an important question regarding this historically contentious relationship between the United States and international law—what are the reasons for American leadership wavering between strong leadership in international law in one historical period and its rejection in favor of isolationism in another?

Scott (2004) traces the growth and evolution of both the realist, or state-dominated, school of international relations and the liberalist, cooperative school, connecting these theoretical approaches to political leaders in the US foreign policy establishment as well as the US position in the world system. Scott's analysis concludes that two factors explain this movement away from a liberalist position to a realist one in regard to international law: a generational change toward believers of realism and state power in positions of influence in the US government, and the rise of the United States as a global superpower. Scott argues that once the Cold War ended and a generational leadership shift occurred, the purpose of international law shifted from an aspirational tool of cooperative international engagement to one of power maintenance.

Legal Realism and International Law

The literature on international legal realism emerges a decade after Scott's (2004) study and is a major source of convergence between the study of international law and international relations. In the last 10 years, scholars have taken up Scott's research question and further analyzed why the US government is moving away from a reliance on public law specialists who adhere to a doctrinal, formal legal, liberal, participative stance on international law and toward its use as an instrumental tool of international power emphasizing a pragmatic, empirical, practice-based use for international cooperation (Bodansky, 2015; Zajec, 2015). From this American perspective, international legal realism is underpinned by longstanding beliefs that domestic politics have a strong impact on the international legal system (Bodansky, 2015) despite the historic separation of normative goals and outcomes from strict formality to the judicial process (Zajec, 2015). Within this literature, a debate continues (Bodansky, 2015; Klabbers, 2015; Shaffer, 2015) as to the sources which define the US purpose for international law, that is, strict adherence to legal doctrine and cooperation or practical use as a predictor of policy outcomes (Shaffer, 2015). The two sides of this argument have a long history in American political thought, connected to concepts of natural law, positivism, and traditions of legal pragmatism as well as international relations theories of liberalist cooperation made famous by President Woodrow Wilson during World War I (Shaffer, 2015).

Jakob Holtermann (2016) discusses this issue further, illustrating two paths this thinking can take: an American variant and a Scandinavian variant. Ontologically speaking, the American variant questions the purpose of international law in a "forward looking legal skepticism"—what decisions are applied in the actual and future practice and use of specific international law? The Scandinavian variant has a focus on "backward looking rule skepticism" and asks, what justifies this rule in the first place? (Holtermann, 2016). Ultimately, the difference between the two variants revolves around a justification for the existence of international law (the Scandinavian variant) and the impact of the policy created through the implementation of the international law (the American variant). The overall impact of international legal realism is to bring the study of international law outside the judges' chambers, as it were, and incorporate both the purpose and impact of such actions into a critical understanding of its effects, a process referred to by Bodansky (2015) as "seeing the entire elephant." In other words, what is most important about international law: adherence to the doctrine of fairness and principles of cooperation or the policy outcomes of such laws?

Domestic Constitutional Issues of International Treaty Incorporation

A corollary to this important discussion concerns the domestic legal mechanism whereby states agree to adopt an international legal treaty as the domestic law of the land—called incorporation. This is important because some states automatically turn any signed international treaty into the domestic law of the land, while others require additional steps of approval for the treaty to become domestic law. Generally, two types of incorporation are practiced by states in relation to international law: automatic standing incorporation (an illustration of legal realism's Scandinavian variant) and ad hoc legislative incorporation (the American variant). In the former, legislation and/or the constitution is worded in an internationalist fashion, automatically adopting as the domestic law of the land any treaty signed by the properly empowered state official, while in the latter, specific legislation is constitutionally required on a case-by-case basis for a signed international treaty to become the domestic law of the land (Cassese, 2005; Shaw, 2014).

States that place great value in international law and collaboration—Western European and Scandinavian, Latin American, and many African and Asian states—frame their constitutional and legal systems to automatically adopt international law and approve of the purpose and inherent use of it, that is, the Scandinavian variant. Countries like the United States and Russia, which aim to put state sovereignty ahead of international legal collaboration, often choose ad hoc legislative incorporation to weigh the usefulness of the policy outcomes generated through international law, that is, the American variant. This general position vis-à-vis international law is often a reflection of the source of power a state chooses to prioritize and doubles back to the earlier discussion of legal realism and the connection between international law and foreign policy. Powerful states may choose to use an ad hoc mechanism to retain some measure of control that they can actually exercise, whereas smaller, weaker states commonly choose international law as their source of power in the absence of the independent ability to do so. Crucially, in states with ad hoc incorporation, this requires an additional legislative step for a signed international legal treaty to become the domestic law of the land.

The United States adheres to ad hoc legislative incorporation, requiring two-thirds approval by the Senate for international law to become the domestic law of the land (Shaw, 2014). President Woodrow Wilson led the creation of the League of Nations and signed the treaty, yet the US Senate refused ratification. Taking a lesson from Wilson's experiences and recognizing this potential legislative pitfall, President Franklin Delano Roosevelt secured passage of the United Nations Charter by obtaining the assent of both parties in advance of treaty negotiation (Owen et al., 1984). In the following period of crisis defined by the Cold War, a consensus developed by which both major US political parties cooperated on many security matters against an outside enemy and minimized the potential impediment of ad hoc legislative incorporation (Hastedt, 2017). While supporting the United States against its Cold War adversary, this consensus did not extend to issues such as the environment, human rights, and other non-security matters during the same period. This shortfall spread to many more issues in the post-Cold War period in which the additional step of domestic political ratification of an international treaty was required, making ratification of international law much more difficult and putting an end to any remaining political consensus.

Domestic Political Gridlock – Putnam's Two-Level Game and the Governmental Politics Model

In this current era of domestic political gridlock, participation in international law has become both an international and domestic political issue, creating not only the need to negotiate signature to a treaty on an international level but also a separate and potentially contentious domestic negotiation to ratify the agreement. The back-and-forth support for and opposition to international law by the United States is better understood using theories which explain the depths of the difficulty in securing agreement in two separate negotiations. Robert Putnam's (1988) article "Diplomacy and Domestic Politics: The Logic of Two-Level Games" and the governmental politics model of domestic foreign policy making by

Graham Allison and Philip Zelikow (1999) offer clarification. Due to the legislative requirement of ad hoc incorporation contained within the US Constitution as well as the ever-changing power structure of and crises in the international system, each international signature and domestic ratification is conducted in a different international context whose conditions change on both levels of negotiation—international and domestic—and whose passage is never assured (Putnam, 1988). In this historically varied climate, beliefs in support of liberal internationalist cooperation in international law come in direct opposition to supporters of policies preferring national independence, sovereignty, and realism in both its international relations and legal varieties (Allison & Zelikow, 1999). The seemingly dramatic swings in support for and in opposition to international law are, therefore, a result of the changing domestic and international contexts in which these decisions to adopt are made and reflect the difficulty in securing both the signature and ratification of treaties on two different levels of negotiation.

Putnam's (1988) two-level game theory enables a researcher to establish the importance of the domestic level of negotiation and leads to the use of Graham Allison and Philip Zelikow's (1999) modeling of US foreign policy making as a theoretical explanation as to how contentious political bargaining over treaty ratification leads to negotiation failure and an uneven participation in international law. Allison and Zelikow's (1999) groundbreaking analysis of the making of foreign policy during the Cuban missile crisis put forward three theoretical models to explain the domestic level of foreign policy making during the Cuban missile crisis, each reflecting the domestic political climate within which the policy was made: the rational actor, organizational behavior, and governmental politics models. The rational actor model frames the foreign policy making process in a context in which the state and its principal policy makers control the process and make decisions based upon a conscious and rational calculation of advantages based on an explicit and internally consistent value system which will result in the achievement of a specific policy goal. The organizational behavior model frames the context as one in which the state's principal policy makers do not steer the policy making process, and ultimate control of foreign policy formation devolves to individual departments or branches of government, which are guided by their own culture, goals, and interpretations of the broader goals of the US government.

Within a contentious climate of domestic political gridlock, in which neither the primary decision-makers nor the departmental bureaucracy control decision-making, lies the third and immediately relevant model developed by Allison and Zelikow (1999)—the governmental politics model. In this theoretical conception, the "governmental actor is neither a unitary agent [as under rational actor] nor a conglomerate of organizations [as under organizational behavior]; rather, it comprises a number of individual players" intent on driving the policy and accomplishing its particular interests in the developing foreign policy (Allison & Zelikow, 1999). In this context of compromise, conflict and competition between actors with diverse interests and unequal influence. Separate institutions sharing power engage in a bargaining game, often in the glaring light of the media and American public, and endeavor to persuade the other actors to develop a policy accomplishing particular goals (Allison & Zelikow, 1999). The reality that this contentious bargaining process produces different winners with varying views on international law helps to explain the uneven participation of the United States in international law.

Methodology

The preceding literature review establishes the importance of the domestic level of negotiation and helps explain why treaties signed on the international level can be difficult to ratify on the domestic level, thus providing some explanation for the uneven participation of the United States in international law. While the executive branch retains an extraordinary amount of power to make foreign policy under the Second Amendment of the US Constitution, this same constitution does not provide automatic standing incorporation of those presidential agreements directly into the domestic law of the land. Additional actors have both direct constitutional power and indirect influence over the final content of the domestic law that emerges from that international legal agreement and, therefore, can affect its ultimate ratification.

This research uses a latent content analysis of peer-reviewed scholarship and historical news articles (Berg & Lune, 2011) which directly reference the contentious process of ratifying international legal agreements in the aftermath of the signature international legal treaty of the early 20th century, that is, the 1919 League of Nations treaty. The following graphic (Table 1) lists specific actors in the foreign policy process to be discussed in this analysis and the main themes revealed by a content analysis of their participation in this process.

Table 1

Actors and their respective issues in foreign policy making

Actor in foreign policy making	Issue of contention/influence over policy making
Legislative branch	• Policy differences • Political partisanship • Posturing to American fears, nativism, isolationism, and xenophobia • Reference to past failures, specifically the League of Nations treaty
Executive branch	• Failure to consult, compromise with Congress during ratification process • Political partisanship • Reference to past failures, specifically the League of Nations treaty
Media and Public Opinion	• Continuing attention and increasing impact on actors in a contentious bargaining process over ratification

These actors and their influence on the domestic context of decision-making in treaty ratification reveal consistent and repeating patterns, illustrating the difficulties inherent in the political process and providing explanations as to why US participation in international law has been and continues to be uneven. Through a thematic content analysis, the following subsections will present evidence of these continuing patterns by important actors as they bargain for the ability to secure or oppose ratification of international legal treaties as part of the US foreign policy making process. While the United States remains deeply involved in international legal agreements, this research provides some explanation as to why that involvement has been, and continues to be, uneven.

The Legislative Branch

While the executive branch—including chief diplomat, head of state, commander in chief of the military, and originator of legislation—retains the lion's share of powers in US foreign policy making under the Second Amendment to the US Constitution, the legislative branch also has the power to make and influence foreign policy (Rosati & Scott, 2013). Well-known as the keeper of the purse, the legislature has the ability to impact foreign policy by authorizing or denying funding for the implementation of a policy, effectively rendering it moot (Rosati & Scott, 2013). The power of the US Senate to ratify or deny ratification of treaties is another important duty of Congress, and one particularly relevant for this study. Whereas the president is the head of state and represents a specific political party, the partisan composition of Congress may increase the difficulty of domestic ratification of a treaty if the Senate majority is of a different party than that of the current president. Even if the party in power in the Senate shares allegiance with the president, international law is often a highly contentious issue and may result in a difficult ratification process. Each of these themes represents additional hurdles for treaty ratification on the domestic level of negotiation, as described in Putnam's (1988) two-level games: (1) policy differences, (2) partisanship and a personal dislike of the president, (3) legislative obstructionism, and (4) comparison to the signature failure of Wilsonian idealism, the League of Nations.

Policy Differences: State Sovereignty, Unilateralism, and Isolationism

The content analysis suggests several policy positions put forward by senators and legislators throughout the 20th century differ with the clear presidential goal to secure ratification of an international treaty recently signed with great fanfare or desired by the executive branch as part of its foreign policy platform. The most prominent of these policy differences include: (1) a desire to maintain state sovereignty, (2) unilateralism of US strength in the international system, and (3) isolationism from the international community. These policy positions are consistent with previously mentioned concepts of American legal realism as they focus on the practical impact of international cooperation and a common domestic American argument that state control of policy necessarily generates positive outcomes for US foreign policy rather than cooperation. In order to bolster opposition to treaty ratification, this content analysis finds a consistent effort to connect the international law currently under consideration to these policy differences and their connection to the failure of the League of Nations treaty, increasing the contentious context of domestic ratification.

Sovereignty. The first of these prominent themes centers around a desire to retain state sovereignty and reduce outside interference in domestic politics as a primary reason to reject treaty ratification. Sovereignty, defined here as a country's ability to govern matters impacted within its borders (Snarr & Snarr, 2016), is an important policy position of all countries, described by preeminent statesman Henry Kissinger (1995) as the "need for balance between international involvement and government authority." Examples of this reluctance to join international legal agreements due to sovereignty concerns include a 15-year delay in joining the International Labor Organization in 1934 (Zelek, 1993) and the needed catalyst of the Great Depression to secure participation in the Bretton Woods and international trade regime after World War II.

The center of this controversy revolves around convincing the government and society that the benefits of international cooperation, on trade for example, outweigh the competing belief that protectionism justified through the concept of sovereign control over trade will yield better results. This is not a new but continuing controversy which has endured for decades, from ratification of the 1948 General Agreement on Tariffs and Trade and its decades-long culmination into the adoption and ratification of the treaty creating the North American Free Trade Agreement in 1993 (Samuelson, 1994; Ackerman & Golove, 1995) and the World Trade Organization in 1995 (Pauwelyn, 2005; Hitt, 2006; Tiefer, 2000) to the actions of the Trump administration in regard to the Trans-Pacific Partnership (Baker, 2017) and the use of tariffs as a sovereign tool to strengthen the economy in areas such as the aluminum and steel industries (Vinicky, 2018).

Unilateralism. The second issue, unilateralism, is often seen in issues of security and is related to the preservation of state sovereignty where it is argued that only "American exceptionalism" and control of the policy and its implementation could secure US interests (Hitt, 2006). Examples of support for policy which adheres to this corollary to concepts of American legal realism include a desire to forego the Reagan era nuclear treaties with the USSR, such as the 1987 Intermediate-Range Nuclear Forces Treaty (Owen et al., 1984; Cranston, 1987); involvement in the Yugoslav wars of the 1990s (Kissinger, 1995; Evans, 1999); the 1997 Chemical Weapons Convention (Clymer, 1997); the 1996 Comprehensive Test Ban Treaty (Apple Jr., 1999); International Criminal Court and its individual tribunals (Strauss, 2011); and a withdrawal from the Anti-Ballistic Missile Treaty in 2002 (Collier, 2003; Rhea, 2011). Frequently justified as a corollary to American superpower status (Cohan, 2001), this research finds that the idea of American exceptionalism is also used as an argument to reject the ratification of international treaties.

Isolationism. Lastly, isolationism is a third policy position which often pervades opposition to international agreements and is an idea which political actors intentionally direct to Americans' fear of outsiders and desire to remain outside of conflict. Examples include repeated decisions to enact military budget cuts and refuse participation in international organizations after international military conflicts such as World War I (Golove, 1999), the United Nations Participation Act after World War II (Schlesinger, 1994), nuclear agreements during the Cold War conflict (Hodgson, 1990), military responses to the conflicts in Yugoslavia (Sorensen, 1995; Frick & Mellon, 1999), international military cooperation in the post-9/11 era (Knowles, 2001; Bederman, 2001; Yoo, 2001; Fisk, 2002), and the 2003 invasion of Iraq (Judis, 1990; Chafets, 2002). These isolationist arguments are equally persuasive in human rights and economic issues, employed with success during the genocides perpetrated against Muslims in Bosnia–Herzegovina (Frick & Mellon, 1999), in support of people with disabilities (Thorsell, 2003), the 1994 Mexican peso crisis (Washington Post, 1995), and the Great Recession of 2008 (Mehta, 2015).

Policy differences are a reasonable and expected element of any domestic negotiation for treaty ratification. Logically, conservative and liberal positions are continuously in opposition in the American domestic political system, and domestic bargaining over policy always consists of political entities pursuing different visions of US foreign policy. The interesting and more surprising conclusions of this research revolve around the somewhat mundane but fundamentally important additional formal legal step of domestic ratification and the repeated focus by the opposition on the failure of the League of Nations as a reason to reject that ratification. The focus on the League of Nations is the foundation of ongoing partisanship which dominates American politics and its inevitable impact on the ratification process required by the US Constitution for the adoption of any international treaty.

Political Partisanship

This spotlight on the failure of Wilsonian liberalism plays directly into a partisan argument consistently put forth by opponents throughout history, that international law fails to advance the interests of the US History indicates that the absence of the US from the League of Nations deprived it of needed strength to maintain peace during the interwar years, and this blame can only be logically placed on the US failure as a whole to find a compromise and join the league, not with blame placed solely on Woodrow Wilson and his political allies (Koplow, 1989; Snarr and Snarr, 2016). Yet, as a tool of political partisanship, this characterization is successfully utilized time and again as an argument to deny the usefulness of international law and Senate rejection of a treaty on partisan grounds.

Many articles have been written about the personal animosity and dislike of the Republican politicians of the time towards the Democrats and Woodrow Wilson as a continued focus of their opposition. This focus on Wilson, as the modern founder of international liberalism, seems logically to escalate the disapproval of a political policy to a specific dislike of an individual politician, especially one held in such regard by the opposing party. Former President Theodore Roosevelt (Cranston, 1987), Senator Henry Cabot Lodge (McMullen 2012; Mehta, 2014), and Elihu Root (Karol, 1988) were contemporary leaders of the opposition who placed blame directly onto Wilson and highlighted their personal dislike of the president. A partisan argument is often used by Senators to reject participation in international law by Republicans during periods in which they control the government but are under pressure to continue international legal cooperation—as in the debate over further nuclear de-escalation treaties under Ronald Reagan in the 1980s (Reston 1982), as well as time periods when the opposing party controls the White House, including the administrations of Bill Clinton (Clymer, 1997) and Barack Obama (McMullen, 2012; Mehta, 2014).

The impact of this recurring problem is aptly described by Robert J. Samuelson of the *New York Times* in the run-up to the 1994 negotiation of the Uruguay round of the World Trade Organization negotiations: "ultimately, at a minimum these [partisan] arguments hamstring the global leadership of the US . . . being less about trade than about how Americans see their role and how other countries view our role" (Samuelson, 1994). Partisanship for its own sake weakens not only policy outcomes but also perceptions of US leadership on the global level, affecting the very authority US leaders intend to secure. The sentiment of partisan opposition is as damaging to US participation today as it was during the international trade negotiations of the 1990s and discussion of League of Nations treaty ratification in 1919, as it arises time and again in the content analysis as an important factor to explain the contentious context of domestic treaty ratification and resulting in the uneven record of US participation in international law.

Ad Hoc Treaty Incorporation and Legislative Obstructionism

The constitutional requirement of treaty ratification by a two-thirds vote of the US Senate enables legislative obstructionism as a method to employ differences in policy as well as political partisanship to complicate the political bargaining and ratification process. These complications often take the form of amendments to the legislation and/or "reservations" which act as legal opt-outs to part of the content of a treaty (Shaw 2014). President Wilson, the Democratic party, and a majority of the public at large supported ratification of the League treaty, yet the addition of numerous amendments (Cranston, 1987; Karol, 1988; Fox, 1999; Mehta, 2014) and constitutional concerns (Johnson, 1993; Rhea, 2011; Ross, 2013) ultimately doomed the ratification vote. These impediments to ratification were based on policy differences, constitutionality, and political partisanship and highlight the often difficult and contentious political context within which a treaty must be approved.

This legal hurdle connects the importance of the domestic level of Putnam's Two-Level games theory to the contentious bargaining among political actors in the competitive context of treaty ratification by the US Senate, policy differences, and partisanship, as well as legal obstructionism.

The President and Executive Branch

Partisanship and obstructionism are, unfortunately, a two-way street and one whose actions extend to the president and executive branch as well. Whether supporting or opposing treaty ratification, presidents throughout history have shown a lack of willingness to compromise, as was true of Wilson during the ratification process for the League. While Wilson's failing health is often considered a reason for his refusal to compromise (Associated Press, 1990; Link, 1990; Evans, 1999; Finkelmeyer, 2009; Berg, 2013), compromise by the president is essential in a competitive bargaining system and within the context of a contentious decision such as treaty ratification. Its absence only adds tension to a characterization of contentious context within Allison and Zelikow's governmental politics model, in which all parties politicize an issue and fail to negotiate a suitable outcome for all sides.

As an indication of the partisanship problem, President Wilson refused to compromise or negotiate with the opposition. In the negotiations in Versailles to end the war and the months before the ratification vote, Wilson acted alone and did not make any effort to work with or include his political opponents and their views in this process (Karol, 1988; Ackerman and Golove, 1995). Once debate began in the Senate, Wilson refused to compromise, adopting a "take it or leave it" stance towards the treaty as a whole (Associate Press, 1990; Clymer, 1997; Evans, 1999; Tiefer, 2000; Mehta, 2014). A clear indication of the importance of this factor in the domestic bargaining process for treaty ratification is the advance agreement secured by President Franklin Delano Roosevelt for ratification of the United Nations Charter; learning from history, and successfully playing the bargaining game among domestic actors in foreign policy, President Roosevelt negotiated a deal in advance so as to weaken the ability of treaty opponents to use the media and public to develop domestic political opposition to weaken support for the deal (Owen, Okita, Brzezinski, 1984).

This refusal to compromise continues in presidents since Wilson and FDR. Despite warnings from the lessons of history and from advisors of the Executive Branch, the chief executive often fails to bargain adequately and in earnest with his political opponents to secure ratification or safeguard an existing treaty. President George W. Bush was steadfast in his adherence to positions in alignment with American legal realism and, rather than negotiate a compromise to an existing treaty, opted instead to remain outside the jurisdiction of the International Criminal Court and Kyoto Protocol and to leave the Anti-Ballistic Missile Treaty altogether (Collier, 2003; Rhea, 2011). In the hyper-partisan climate of the Obama administration, despite the President's recognition of the uphill battle for ratification (Finkelmeyer, 2009), an outward desire to negotiate was neither enough nor successful, and opportunities for Senate ratification to the 2012 United Nations Disability Treaty (McMullen, 2012) and Paris Climate Accord failed (Mehta, 2014; Conn, 2017). President Trump continued this pattern of partisanship, intentionally seeking out the legislative accomplishments of his predecessor and eliminating them on partisan grounds, often with negligible grounding in principles of American legal realism or the national interests of US foreign policy (McGrane, 2017). The evidence indicates that presidents often do not recognize the importance of compromise to secure ratification of an international treaty or seek purely partisan gain after the realization that domestic political gridlock renders a successful negotiation too difficult.

The Media and Public Opinion

Then as now, public opinion often indicates that American majorities favor the ratification of many international treaties on the domestic level, so why does the US have this uneven record of participation in international law? Again the answer hinges on the multiple step process of international agreement and the domestic ratification made increasingly contentious through media coverage and public attention. An analysis of the failure of ratification for the League of Nations Treaty in the years that followed showed clear majorities of Americans in favor of ratification (Berdahl 1929; Cranston, 1987; Frick and Mellon, 1999; Levins, 2013). Yet it was the contentious domestic political bargaining and policy differences which produced partisan gridlock rather than compromise and negated majoritarian support for ratification by the American population. The constitutional requirement of a two-thirds vote on ratification is an often significantly higher threshold than the consistent majorities of Americans in support for a treaty so frequently emboldens the opposition to vote the agreement down (Reston, 1982; Fox, 1999; Knowles, 2001; Westell, 2002).

These intense battles between prominent political actors in the contentious bargaining game over treaty ratification as US foreign policy play out in the full view of the public with the assistance of the mainstream media. The continuous attention of the media develops public opinion, which has an increasingly important impact on the making of US foreign policy. This increased attention also has the perhaps unintended effect of ratcheting up both the intensity of the argument and the potential divides among the legislative and executive branches and all actors involved in treaty ratification. The media spotlight on display during the period of negotiation of the League treaty now shines exponentially brighter as technological developments have progressed from radio to television, cable TV to 24-hour internet and social media; with these technologies, the power of the media and public opinion is a stronger actor in US foreign policy (Hastedt, 2017).

The ability to reject ratification of international treaties is further advanced through the attention given by the media to common, negative themes which dampen public enthusiasm for international law: isolationism, xenophobia, nativism and nationalism. Whether considering 1919 or 2019, an observer of politics recognizes these common themes. During the ratification negotiations, opponents of the League of Nations Treaty used these themes to generate fear and opposition, whether or not grounded in evidence (Johnson, 1993; Fox, 1999; Frick and Mellon, 1999; Bederman, 2001; Westell, 2002; Conn, 2017). In the modern era of ever-present social media and 24-hour cable news, the prospects for domestic ratification of international law become increasingly slim and provide additional explanation of the uneven involvement of the US in international law.

Conclusions

This paper seeks to analyze the complicated and uneven relationship between the US and international law. It surveys the complementary literatures concerning the domestic adoption of international law, international relations theory, and US foreign policy making and uses a latent content analysis of peer-reviewed studies and contemporary news articles to explain this uneasy relationship. The intention of this research was to contribute to existing scholarship on the issue and provide additional explanation as to why the US has historically had an uneven relationship with the international legal cooperation that played such a positive role in its emergence as a global power on the world stage.

This research suggests several interesting conclusions which contribute to scholarship on the topic. First, this research concludes that liberalist,

Wilsonian support for international law continues to exist alongside a separate undercurrent of distrust, isolationism, and skepticism over the practical use of international law as a tool which can achieve the goals of US foreign policy. The policy differences between Wilsonian idealism and American legal realism are borne out through the thematic support for protecting sovereignty, unilateralism, and isolationist threads which have consistently been part of the domestic political context of these political issues since World War I and the ratification controversy of the League of Nations Treaty, and well before then. The content analysis provided numerous examples throughout the decades of the 20th and 21st century of international legal issues in which the policy differences of members of the US Congress play a part and which contribute to explanations of the failure to adopt international law on the domestic level.

Second, the US constitutional requirement of Senate ratification of any international treaty signed by the president creates an additional legal step for such an agreement to become domestic law, a process which often becomes mired in the contentious political context which Graham Allison and Philip Zelikow identify as the governmental politics model of US foreign policy making. This bargaining process more than occasionally ends with the failure of ratification of the international legal treaty, illustrating the importance of the domestic level of negotiation as put forward in Robert Putnam's Two-Level Games. This legal step required in the US creates an additional hurdle in the process of adopting international law in this country, thereby providing additional explanation for the uneven involvement of America in international law.

Third, and perhaps most surprisingly, the failure of the League of Nations Treaty in 1919 is the historical event most used as an illustration in opposition to international law during the ratification process. As the signature effort of the US president most associated with the view of liberalist cooperation, the failure of the League of Nations continues to be a rallying cry in opposition to international law. The content analysis indicates strong support for the finding that the League of Nations failure and the rejection of its treaty by Congress are proof why international law does not advance American interests. Partisan, contentious political negotiations among increasing numbers of powerful actors in the domestic ratification process are an additional hurdle which treaties signed on the international level must surpass in order to become part of the law of the land. This additional step is the arena for contentious political bargaining, a failure of negotiation casting a shadow over consistent support for international law by the US.

References

Ackerman, B., & D. Golove. (1995). Is NAFTA constitutional? *108 Harvard Law Review 801-929.*

Allison, G., & P. Zelikow. (1999). *Essence of decision: Explaining the Cuban missile crisis*, London, Pearson.

Apple, R.W. Jr. (1995). Defeat of a treaty: news analysis; the G.O.P. torpedo. *The New York Times.*

Associated Press. (1990). Historian says illness sapped Wilson's skill, *pp. 33A.*

Baker, P. (2017). Trump abandons trans-pacific partnership, Obama's signature trade deal. *The Associated Press, pp. A1.*

Bass, G.J. (2014). Last country standing. *The New York Times, pp. 20.*

Bederman, D.J. (2001). National security: globalization, international law and United States foreign policy. *50 Emory Law Journal 717.*

Berdahl, C.A. (1929). The United States and the League of Nations. *Michigan Law Review, 27(6), 607-636.*

Berg, B.L., & H. Lune. (2011). *Qualitative research methods for the social sciences*, London, Pearson Longman.

Bodansky, D. (2015). Legal realism and its discontents. *Leiden Journal of International Law, 28,* 267–281.

Cassese, A. (2005). *International law.* Cary, Oxford University Press.

Chaffets, Z. (2002). Bush's double ultimatum: one to Saddam and one to the United Nations. *Daily News (New York), pp. 43.*

Clymer, A. (1997). The chemical arms treaty: the overview; senate approves pact on chemical weapons after Lott opens way. *The New York Times, pp. 1.*

Cohan, C.C. (2001). International mavericks: a comparative analysis of selected human rights and foreign policy issues in Iran and the United States. *33 George Washington International Law Review, 197.*

Collier, M.M. (2003). Between empire and community: the United States and multilateralism 2001-2003: A mid-term assessment: minimum public order: the Bush administration's reaction to September 11: a multilateral voice or a multilateral veil? *21 Berkeley Journal of International Law, 715.*

Conn, S. (2017). Isolationism is another word for avoiding inconvenient truths. *The Korea Herald.*

Cranston, A. (1987). A debacle like the League of Nations? *The New York Times, pp. 23.*

Evans, H. (1999). Why this is a peace of history; start the nit-picking tomorrow; today be glad the US went to war. *The Guardian (London), pp. 20.*

Finkelmeyer, Todd. (2009). Wilson's legacy explored in UW prof's book. *The Capital Times (Madison, Wisconsin).*

Fisk, R. (2002). How the League of Nations ended up as debris. *The Independent (London), pp. 19.*

Fox, A.W. (1999). The League sinks. *The Washington Post, pp. C16.*

Golove, D. (1999). Part III: from Versailles to San Francisco: the revolutionary transformation of the war powers. *70 University of Colorado Law Review 1491.*

Hastedt, G.P. (2017). *American foreign policy: past, present and future.* Lanham, Rowman & Littlefield.

Hitt, G. (2006). WTO at crossroads as failed talks cloud its future; risks fate of the League of Nations: a failed experiment in global governance. *The Globe and Mail (Canada), pp. B6.*

Hodgson, G. (1990). Swords that will not melt. *The Independent (London), pp. 17.*

Holtermann, J.V.H. (2016). Getting real or staying positive: legal realism(s), legal positivism and the prospects of naturalism in jurisprudence. *Ratio Juris, 29, 535–555.*

Johnson, R.D. (1993). Article XI in the debate on the United States' rejection of the League of Nations. *15 International History Review, 3, 502-524.*

Judis, J.B. (1990). George Bush, meet Woodrow Wilson. *The New York Times, pp. 21.*

Karol, J. (1988). Wilson brought the League of Nations debacle on himself. *The New York Times, pp. 26.*

Klabbers, J. (2015). Whatever happened to Gramsci? Some reflections on new legal realism. *Leiden Journal of International Law 28, 469–478.*

Knowles, R. (2001). Starbucks and the new federalism: the court's answer to globalization. *95 Northwestern University Law Review 735.*

Koplow, D.A. (1989). Arms control treaty reinterpretation: article: constitutional bait and switch: executive reinterpretation of arms control treaties. *137 University of Pennsylvania Law Review 1353.*

Koring, P. (2003). Rumsfeld conjures up Mackenzie King quote. *The Globe and Mail (Canada), pp. A13.*

Levins, H. (2013). Book review: 'Wilson': A serious man for serious matters.

The influential president was involved in landmark laws and brought the U.S. into world war I, by A. Scott Berg. *St. Louis Post-Dispatch (Missouri).*

McGrane, V. (2017). Trump's greatest mission: erasing Obama's legacy. *The Globe.*

McMullen, D.L. (2012). Republican hubris, then as now. *Tampa Bay Times, pp. 11A.*

Mehta, H. (2014). What Obama could pick up from Wilson; one lesson would be in how not to build a new world order the way Wilson did in 1918-1919; another would be in how not to manage his relationships with republicans. *The Business Times Singapore.*

Mehta, H. (2015). Many US policymakers tilt towards trade protectionism; the effects of the crash of 2008 are still being felt in 2015, with growing evidence that protectionists in Congress may block two major free trade pacts. *The Business Times Singapore.*

Owen, D., & O. Saburo, & Z. Brzezinkski (1984). Exploiting Summitry. *The New York Times, pp. 21.*

Pauwelyn, J. (2005). The transformation of world trade. *104 Michigan Law Review 1.*

Putnam, R.D. (1988). Diplomacy and domestic politics: the logic of two-level games. *International Organization, 42, 427-460.*

Reston, J. (1982). Washington; the politics of A-bombs. *The New York Times, pp. 21.*

Rhea, H.M. (2011). Human rights & U.S. standing under the Obama administration: Paris 1919 and Rome 1998: different treaties, different presidents, different senates, and the same dilemma. *20 Transnational Law & Contemporary Problems 411.*

Rosati, J.A. & J.M. Scott. (2013). *The politics of United States foreign policy.* Independence, Thomson Wadsworth.

Ross, W.G. (2013). Constitutional issues involving the controversy over American membership in the League of Nations, 1918-1920. *53 American Journal of Legal History 1.*

Rubenfeld, J. (2004). Commentary: unilateralism and constitutionalism. *79 New York University Law Review 1971.*

Samuelson, R.J. (1994). Remember the League of Nations. *The Washington Post, pp. A19.*

Scott, S.V. (2004). Is there room for international law in realpolitik? Accounting for the US 'attitude' towards international law. *Review of International Studies, 30, 71-88.*

Schlesinger, Arthur. (1994). America's 50-year century. *The Times.*

Shaffer, G. (2015). International legal theory, international law and its methodology: the new realist approach to international law. *Leiden Journal of International Law*, 28, 189–210.

Shaw, M.N. (2014). *International Law.* Cambridge, Cambridge University Press.

Sheridan, Michael. (1995). America relives an old debate – and the war grinds on. *The Independent (London), pp. 13.*

Snarr, M.T., & D.N. Snarr (Eds.). (2016) *Introducing global issues*, Boulder, Lynne Rienner.

Sorensen, T.C. (1995). The star spangled shrug; is America shirking its leadership role? *The Washington Post, pp. C01.*

Strauss, A. (2011). Cutting the Gordian knot: how and why the United Nations should vest the International Court of Justice with referral jurisdiction. *44 Cornell International Law Journal 603.*

Swaine, E.T. (2001). The local law of global antitrust. *43 William & Mary Law Review 627.*

Thorsell, W. (2003). Who governs: the pretense of a balance of power. *The Globe and Mail (Canada), pp. A17.*

Tiefer, C. (2000). Adjusting sovereignty: contemporary congressional-executive controversies about international organizations." *35 Texas International Law Journal, 239.*

Vinicky, A. (2018). Precious Meddling: The Impact of Trump's Tariffs on Steel. *Chicago Tonight.*

Washington Post. (1995) Mexico's ineffectual neighbor to the north, *pp. A27.*

Westell, A. (2002). Another League of Nations. *The Globe and Mail (Canada), pp. A19.*

Yoo, J.C. (2001). Laws as treaties? The constitutionality of congressional-executive agreements. *99 Michigan Law Review 757.*

Zasloff, J. (2003). Law and the shaping of American foreign policy: from the gilded age to the new era. *78 New York University Law Review 239.*

Zajec, O. (2015). Legal realism and international realism in the United States during the interwar period: Neglected reformist convergences between political science and law. *Revue Française de Science Politique (English Edition), 65(5-6), 51-70.*

Zelek, M.E. (1993). Book review: James Michael Zimmerman, extraterritorial employment standards of the United States: the regulation of the overseas workplace, (New York: Greenwood Publishing Group, Inc., 1992). *14 Comparative Labor Law 514.*

8

Cruelty or Military Necessity: Poison Gas and U.S. Security in WWI

Thomas I. Faith
U.S. Department of State

Abstract

The First World War made chemical weapons a security issue for the United States. Poison warfare aroused controversy since ancient civilization, but World War I involved the United States in the legal and ethical debates surrounding the practice. The United States emerged from the war with concerns about how chemical weapons might affect domestic and international security in the future, and its policy makers began working toward more comprehensive, enforceable arms control measures.

Keywords: Arms Control; Chemical Warfare; Foreign Policy; Geneva Protocol; World War I

Cruelty or Military Necessity: Poison Gas and U.S. Security in WWI

The First World War made chemical weapons a security issue for the United States. Poison warfare has aroused controversy since ancient civilization, but World War I involved the United States in the legal and ethical debates surrounding the practice. The United States emerged from the war with concerns about how chemical weapons might affect domestic and international security in the future, and its policy makers began working toward more comprehensive, enforceable arms control measures.

Poison Warfare Before World War I

The First World War represented the earliest widespread military use of chemical weapons in the modern era, but poison weapons have been used since antiquity. To understand how the use of chemical weapons in the Great War came to be interpreted in the war's aftermath, gas warfare should be viewed in context with ancient poison weapons. The methodological predecessors of modern chemical weapons provided the moral basis in which those weapons were understood during World War I.

References to poisoned weapons appear in some of the earliest texts—in Homer's *Odyssey*, for example, where Odysseus poisons his arrows with hellebore (Mayor, 2003, pp. 32–33). Thucydides describes the use of toxic smoke during the siege of Plataea in 428 BC in his *History of the Peloponnesian War*. The circa 150 AD Hindu Laws of Manu, perhaps the oldest code of conduct to prohibit chemical warfare, forbade the use of weapons tipped with poison. In contrast, the *Arthashastra*, written also in India but several centuries later, contains recipes for poison weapons suggested for military use. While poison has been used as a military weapon since recorded history, the balance of literature about its use indicates that it never enjoyed categorical acceptance as a tool of war.

Moral debates over the use of poison in warfare were kept alive, at least in part, by political elites of centralized states. It is likely that elites feared poison more than other types of weapons because it was a stealth weapon that could be effectively wielded by anyone. A blade in the hands of an untrained novice became far more deadly if it was tipped with poison, and poisoned water or food was arguably a greater threat than enemy armies to both leaders and populations (Price, 1997, pp. 24–25). Accordingly, harsh punishments were prescribed for those who used poison, such as in England where Henry VIII in 1531 ordered "poisoners to be boiled to death" (Roberts, 1915, p. 56). Poison became established as a weapon associated with weakness and cowardice, even as it continued to be employed. For example, in the 15th century, Leonardo da Vinci designed a projectile filled with a toxic powder for use in war. In 1672, Christoph Bernhard von Galen, the Bishop of Münster, used toxic explosive devices at the siege of the city of Groningen (Tuorinsky, 2008, p. 11).

Debates about the use of poison weapons continued through the development of poison gas weapons in the 19th and 20th centuries. In 1812, Thomas Cochrane, a British naval officer, submitted a formal plan to drive Napoleon's soldiers from Toulon, Flushing, and other ports by burning ships loaded with sulfur and coal to create noxious clouds of sulfur dioxide (Miles, 1970, pp. 297–298). The plan was reviewed and rejected by a committee on the grounds that winds and tides were too unpredictable to make the idea practical. When Cochrane reintroduced the idea in 1846, a committee rejected it again, this time on the grounds that the use of poison smoke violated the rules of warfare and could be copied and used against Britain by some future enemy. Cochrane nevertheless reintroduced the idea at least twice more during the Crimean War, though it was never adopted.

Cochrane was not the only Briton to suggest using toxic smoke against the Russians in Crimea. Lyon Playfair, a British chemist, devised a cannon shell filled with cacodyl cyanide, a foul-smelling toxic compound (Miles, 1970, p. 299). The shell was designed to be made of brittle metal so that it would shatter on impact and spread the chemical. Playfair presumably arrived at his poison gas shell design independently of Cochrane, because the latter's toxic smoke proposals did not become public until the 1890s. The British War Department rejected Playfair's poison gas weapon as well, deeming it a violation of the rules of war.

During the US Civil War, the use of poison was expressly prohibited by President Abraham Lincoln and the War Department on April 24, 1863, by General Orders No. 100, also known as the Lieber Code. Article 16 of these instructions for the conduct of US armies in the field states, "Military necessity does not admit of cruelty—that is, the infliction of suffering for the sake of suffering or for revenge, nor of maiming or wounding except in fight, nor of torture to extort confessions. It does not admit of the use of poison in any way, nor of the wanton devastation of a district" (*General Orders*, 1864, No. 100). Further, Article 70 of the Lieber Code reads, "The use of poison in any manner, be it to poison wells, or food, or arms, is wholly excluded from modern warfare. He that uses it puts himself out of the pale of the law and usages of war" (*General Orders*, 1864, No. 100).

Despite this sentiment, there are a handful of documented proposals to use poison gas weapons during the US Civil War. On April 5, 1862, John Doughty, a New York schoolteacher, sent a letter to the War Department describing a design for a chlorine-gas-filled cannon shell (Miles, 1970, pp. 300–302). The War Department took no known action on the letter, and it was likely forgotten amid the torrent of new weapons ideas flooding into their offices at that time. A well-known geologist and agricultural chemist, Forrest Shepherd, wrote President Abraham Lincoln in 1864 to propose that hydrogen chloride be used against the Confederates during the siege of Petersburg. As with Doughty's proposal, Forrest's seems also to have been ignored. A Washington, DC, chemist, William C. Tilden, also proposed the use of a type of chemical weapon, but nothing is known about the circumstances or its design (Miles, 1970, p. 303). According to Tilden's 1905 obituary in the *Evening Star*, General Ulysses S. Grant discouraged Tilden's proposal because "such a terrific agency for destroying human life should not be permitted to come into use by the civilized nations of the world" (Miles, 1970, p. 303).

The International Declaration concerning the Laws and Customs of War, drafted in Brussels on August 27, 1874, was the first attempt to codify an international prohibition against poison weapons in the 19th century. Article 13a of this Brussels Declaration specified that the use of poison or poisoned weapons was "strictly forbidden" as a means of war (Higgins, 1904, p. 27). While the language of the article prohibited all poison weapons, gaseous weapons were not specifically indicated, and it is unclear whether the conference delegates intended them to be included in the prohibition. The Brussels Declaration never entered into force, but this draft was used by delegates at the subsequent International Peace Conference at The Hague, where poison gas projectiles were specifically discussed.

The International Peace Conference took place in 1899 at the invitation of the government of Russia. On May 31, at the third meeting of the subcommission on naval matters, Russian delegate Captain Scheine stated that his government had "instructed him to make a proposition concerning the prohibition of the putting into use of any new kind of explosive, the invention of which seems possible" (Scott, 1920, p. 366). He explained, "It is a question of prohibiting the use of projectiles loaded with explosives which spread asphyxiating and deleterious gases" (Scott, 1920, p. 366). After Scheine clarified for the committee members that he was speaking only of projectiles whose purpose was to spread gas, not projectiles that emitted gas incidentally, Captain Alfred Thayer Mahan of the US delegation declared himself opposed to such a prohibition.

Mahan admitted that he had "not made a special study" of this type of weapon, but nevertheless expressed his opinion that the use of gas projectiles "can not be considered as being a means which is prohibited on the same ground as the poisoning of waters" (Scott, 1920, p. 366). He continued, "Such projectiles might even be considered as more humane than those which kill or cripple in a much more cruel manner by tearing the body with pieces of metal" (Scott, 1920, p. 366). Mahan opined that the suggested projectile might prove effective as a weapon, and that its use "would involve neither useless cruelty nor bad faith, as exists in the case of poisoning waters" (Scott, 1920, p. 366).

Following some discussion over whether the subcommission had jurisdiction over this issue, the delegates continued to speak about whether poison gas projectiles should be prohibited (Scott, 1920, p. 366). Chamberlain Bille, the head of the delegation from Denmark, expressed support for Scheine's proposal and speculated that gas projectiles could pose a greater risk to inhabitants of besieged cities than ordinary projectiles. Immediately afterward, the president of the subcommission conducted a vote on the proposal. Mahan, speaking for the US delegation, was the only vote against the prohibition.

The head of the US delegation, Andrew Dickson White, attempted to convince Mahan to reverse his opposition and support the prohibition, but Mahan refused and asked that his explanation be entered into the record (Davis, 1962, p. 119):

> 1. The objection that a warlike device is barbarous has always been made against new weapons, which have nevertheless eventually been adopted.

In the middle ages firearms were accused of being cruel; later on an attack was made against shells, and still more recently (the author remembers this very well) against torpedoes.

It does not seem demonstrated to him that projectiles filled with asphyxiating gases are inhuman and uselessly cruel devices, and that they would not produce a decisive result.

2. He is the representative of a nation which is actuated by a keen desire to render war more humane, but which may be called upon to make war, and it is therefore necessary not to deprive one's self, by means of hastily adopted resolutions of means which might later on be usefully employed. (Scott, 1920, pp. 366–367)

The final text, ratified on July 29, 1899, read that "The Contracting Powers agree to abstain from the use of projectiles the object of which is the diffusion of asphyxiating or deleterious gases" (Higgins, 1904, p. 78). In force at the outbreak of the First World War, all the principal belligerents were signatories, with the exception of the United States. The Hague Declaration prohibiting poison gas projectiles was the most significant prewar development regarding chemical arms limitation, and it framed the debates surrounding chemical warfare during World War I and its aftermath.

Poison Warfare in World War I

Poison gas weapons appeared on the battlefields of World War I before the United States' formal involvement (Tuorinsky, 2008, pp. 12–14). In late 1914, as tactical stalemate prevailed and the warring nations began to experience weapons shortages, Britain, France, and Germany began to consider using toxic gas or chemical irritants in combat operations along with other alternatives to conventional weapons. The French used rifle grenades filled with tear gas against the advancing German army in August 1914, but the weapons were so ineffective it is possible that the Germans were unaware of their use until later. In October 1914, the Germans fired artillery shells filled with tear gas and other chemical irritants against the British at Neuve-Chapelle and then against the Russians at Bolimów in January 1915, but in both cases the chemical agents in the weapons proved ineffective. In each instance, the perpetrators of these chemical attacks did not feel that their actions had violated The Hague Declaration because the gasses employed were nonlethal.

The infamous gas attack at Ypres on April 22, 1915, was the first effective gas attack of World War I. Under the direction of chemist Fritz Haber, German Pioneer Regiment 35 conducted a carefully planned chemical assault against British, Canadian, French, and Algerian soldiers in defensive positions across no-man's-land. The Germans released chlorine gas from storage cylinders along the front lines and allowed the wind to sweep the gas west toward its anticipated targets. The poison gas proved devastating for the unprepared defenders, and, in the aftermath, its use was denounced as a violation of the rules of war. British commander Sir John French in a July 12 *New York Times* article called the use of poison gas a "cynical and barbarous disregard of the well-known usages of civilized war" ("French condemns the use of gas," 1915). His characterization of poison weapons as barbarous and uncivilized was consistent with the opprobrium historically associated with such methods of warfare, even if the fact that projectiles were not used meant that the attack was not technically a violation of The Hague Declaration.

As it was in antiquity, even as poison-weapons use was denounced in the weeks and months that followed, poison gas weapons gradually became a salient feature of the fighting on the Western Front. The British retaliated in kind against the Germans at Loos on September 24, 1915, using Haber's method of releasing the gas and allowing it to drift with the prevailing wind, but through 1915, 1916, and 1917, the warring powers developed chemical artillery shells and other methods of delivering poison gas that were less dependent on weather conditions and were clear violations of The Hague Declaration.

When the United States sent soldiers to France after declaring war in 1917, it sent them into a war where poison gas was already being used extensively, so there was no substantive debate about whether gas weapons should be used in combat. The hastily formed Chemical Warfare Service of the US Army committed itself to fighting the chemical war as ably as resources would allow. The 30th Engineers were rechristened the 1st Gas Regiment and staffed with warrior-chemists to conduct chemical warfare operations for the American Expeditionary Forces. They launched their first poison gas attack against the Germans at La Ferme Saint Marie on the evening of June 18, 1918, and US artillery units utilized gas shells in combat operations along with their French and British counterparts.

As they worked to produce chemical munitions and refine their use, the officers of the US Army Chemical Warfare Service convinced themselves that the taboo against poison weapons was fading to insignificance. The widespread use of poisonous gases as weapons in World War I clearly meant The Hague Declaration had failed its first test. The signatory nations had not refrained from violating the spirit or the letter of the 1899 agreement, and Mahan's prediction that poison gas weapons might be usefully employed seemed justified in hindsight.

But this was a narrow perspective: the outbreak of chemical warfare had not erased the poison-weapons taboo any more than instances of poison-weapons use had in antiquity. On the contrary, poison gas continued to be denounced as a violation of military norms by the belligerents even as it was widely employed. Moreover, no nation deliberately used poison gas against civilians despite opportunities to do so. Military organizations sanctioned long range artillery and aerial bombardment of civilian areas and unrestricted submarine warfare against passenger ships, but they refused to use gas against noncombatants in a similar fashion.

Inevitably, some civilians were affected by poison gas. In his memoirs, British commander Charles H. Foulkes wrote about the effects of poison gas on the French city of Lille, observing "vegetation and salads in the gardens at Lille had often been bleached by the gas clouds, and the inhabitants were constantly being warned of their approach by the ringing of bells, when they took refuge in the top rooms of their houses and closed all the windows and doors" (Foulkes, 1934, p. 296). No army made a systematic attempt to record noncombatant gas casualties, so information on civilians injured or killed by poison gas is episodic. The highest number of civilian gas casualties from a single incident that is known to have occurred happened in Armentiéres on July 28–29, 1917. Remnants of the town's population were caught in a mustard gas bombardment, and they sheltered in cellars where, unfortunately, the substance tended to concentrate. The mustard gas caused 675 civilian casualties, of which 86 were deaths (Thomas, 1985, pp. 20–21). Civilians near the front lines were subject to gas attacks; however, they never seemed to be specifically targeted.

The potential for poison gas weapons to be used against cities caused considerable dread on the home front, even as far away from the fighting as the United States. On August 12, 1918, the commander of the Smith Island Coast Guard station in North Carolina raised a national alarm when he claimed that a German U-boat had attempted to attack Wilmington with mustard gas. A noxious oily substance had washed ashore there, killing a brood of chickens and causing three members of the Coast Guard station and three people at a nearby lighthouse to pass out. The *Washington Post*, *New York Times*, and *Chicago Daily Tribune* reported the Coast Guard's allegation that the incident was a poison gas attack and speculated that it was the beginning of a new U-boat offensive against the Atlantic coast (Fox, 1918; "Lighthouse Crew Gassed," 1918; "U-Boat Attempts to Gas US City," 1918). The substance that caused the injuries and chicken deaths, however, was most likely released from the damaged batteries of a U-boat that had been depth charged by a US destroyer off the mid-Atlantic coast at approximately the same time. The chemical war against US ports never materialized, and the First World War ended without a single authentic poison gas attack on US soil.

Americans in war material manufacturing plants were sometimes exposed to poison gas. Working with mustard gas was particularly hazardous (Faith, 2012). At Edgewood Arsenal, the Chemical Warfare Service's central manufacturing complex, there were 922 injuries and three deaths from June to December 1918 in all the poison gas plants combined. The majority of these injuries, 673, and one of the deaths occurred among workers at the mustard gas factory (Fries, West, 1921, p. 59). The Dow Chemical Company in Midland, Michigan, noted that at least two minor injuries occurred every day in their mustard gas plant, and the steady number of casualties resulted in the establishment of a hospital on the premises (Steen, 2014, p. 107). A June 1918 accident at Dow caused five severe mustard gas burns and two deaths. When a similar plant was established to manufacture toxic lewisite at Willoughby, Ohio, the Chemical Warfare Service estimated that 10% to 25% of the workers would be injured at a given time and recommended that the workforce be 50% larger to accommodate the attrition (Steen, 2014, p. 109). War workers were not the only persons at risk from accidents at poison gas plants. On August 3, 1918, an accident occurred at the American University experiment station in Washington, DC, that released poison gas into a nearby neighborhood ("N.B. Scott 'Gassed,'" 1918). The toxic cloud injured Senator Nathan B. Scott, who was sitting on the porch of his home and could not reach safety indoors before he succumbed.

The irony is that while the gassing of civilians was not widespread or purposeful, the possibility that noncombatants might be targeted by chemical attacks in future wars became the focus of chemical arms limitations talks after World War I. Despite The Hague Declaration's failure to prevent poison gas warfare from occurring, governments persisted in attempting to prohibit chemical warfare in subsequent international agreements. Those who sought to prevent the future use of chemical weapons shared the belief that such international agreements would eventually be successful, and they thought they were critical for the protection of noncombatants.

Postwar Arms Control

Supporters of chemical warfare believed that US security interests had fundamentally changed at the end of the World War I. For them, the failure of The Hague Declaration to prevent chemical warfare was an indication that chemical weapons would inevitably be used in future wars—and that the United States would have to prepare accordingly. A captain in the Chemical Warfare Service described the "unlimited chemical warfare" that he predicted would characterize future armed conflicts:

> Gas will be the greatest aid to the belligerent whose resources in gas production are greatest. At the signing of the Armistice in 1918 the United States was easily the leader in this respect. In view of her resources in raw material and financial power it would be to the advantage of the United States to use gas against any enemy or combination of enemies. Her vast productive capacity would insure [sic] that her troops could engage in unlimited Chemical Warfare day and night, summer and winter until the end of the war. (Gempel, 1925)

The future of chemical warfare began to be decided at Versailles in 1919. Prognostications by chemical warfare supporters aside, Germany was responsible for the devastating chlorine gas attack at Ypres in 1915, and that event featured prominently in anti-German propaganda throughout World War I as evidence of the German military's brutality. It was natural for the representatives of the victor nations at Versailles to consider restrictions on Germany's ability to wage chemical warfare. Additionally, Germany's chemical industry was seen as a threat to manufacturing in Britain, France, and the United States. In consultation with Secretary of State Robert Lansing and the French, the British proposed the language that became Article 171 of the treaty of peace with Germany (Spiers, 1986, p. 35):

> The use of asphyxiating, poisonous or other gases and all analogous liquids, materials or devices being prohibited, their manufacture and importation are strictly forbidden in Germany.
>
> The same applies to materials specially intended for the manufacture, storage and use of the said products or devices. (Boserup, 1973, p. 153)

The language of Article 171 differed from prior poison weapons agreements, such as The Hague Declaration, because it reflected an appreciation for the relationship between restrictions on manufacturing and prohibition of use. Whereas previous agreements outlawed the use of poison weapons, Article 171 also addressed manufacture, importation, and storage. After World War I came to a close, some of the technical experts who aided the war effort assumed roles in policy making. Chemists who had waged chemical warfare advocated for a variety of domestic and international policies based on their wartime experiences and influenced the next generation of arms control measures.

Some of the most well-circulated ideas about chemical weapons limitations were written by Victor Lefebure, a British chemical-warfare officer. American policy makers were widely exposed to his writings. Charles H. Herty, president of the American Chemical Society and the Organic Chemical Manufacturers Association, gave a copy of Lefebure's book, *The Riddle of the Rhine: Chemical Strategy in Peace and War*, to Elihu Root, former secretary of war, secretary of state, and president of the Carnegie Endowment for International Peace, for use during the Washington Arms Limitation Conference in 1921 (Herty, 1921). A copy of a Lefebure article on chemical disarmament was provided to Secretary of State Charles Evans Hughes by the US Army Chemical Warfare Service for the Washington Conference as well (Lefebure, 1921). In that article, Lefebure criticized those who "think it sufficient to issue an edict against the use of poison gas, not realizing that this alone is absolutely futile as an effective measure" (Lefebure, 1921, p. 6). He advocated a program of inspections and controls designed to cause "the serious reduction of the producing capacity of the German dye monopoly" (Lefebure, 1921, p. 7).

Chemical manufacturers and policy makers had motives to make such recommendations beyond the interests of international security: they had powerful economic reasons for doing so. Breaking the German dye monopoly or otherwise limiting Germany's chemical industry would allow competing manufacturers in the United Kingdom, France, and the United States to fill global demand for dyes and other chemical products. Lefebure and other chemical industry supporters did not advocate chemical weapons arms control measures outside of Germany after World War I. On the contrary, they promoted policies designed to limit Germany's chemical manufacturing by expressly arguing that such policies would give the Allies military advantages if another war occurred in the future. Nevertheless, industry experts from the United States and Allied nations challenged policy makers to think about methods of prohibiting chemical weapons in wartime by limiting importation, manufacturing ability, and stockpiles in peacetime.

While this trend led to agreements with inspections regimens and other enforceable provisions in the latter half of the 20th century, the victors of World War I were far less willing to impose the same restrictions on themselves that they agreed to impose on Germany after the Armistice. At the Washington Arms Limitation Conference in 1921, Secretary of State Charles Evans Hughes proposed the adoption of an international prohibition against poison gas, but without reference to the limits on manufacturing, importation, and storage included in the Treaty of Peace with Germany. Hughes stated that a declaration against poison gas weapons was needed "such as that a whole city should not be asphyxiated on any pretext whatsoever, or that women and children—no part of the armed forces—should not be killed by use of bombs against civilians" (Memorandum, n.d., p. 5).

Amos A. Fries, chief of the Chemical Warfare Service, served as a member of the US delegation at the Washington Conference, and he strenuously objected to Hughes's proposal. Like other chemical warfare officers, Fries believed that poison gas weapons were effective tools of war and that the United States should oppose international agreements designed to prohibit them as Alfred Thayer Mahan had in 1899. Fries argued that chemical warfare was neither cruel nor barbaric, and, to the specific point about the threat poison gas posed to civilians, he stated, "it can be limited in its use against non-combatants only to the extent that other weapons of warfare are limited" (Fries, 1921, p. 6).

Other US delegates at the Washington Conference considered what Fries and his like-minded chemical warfare experts had to say, particularly on the subject of how poison gas might be used against civilians, but they came to different conclusions. The Subcommittee on New Agencies of Warfare was one of the groups at the Washington Conference that reviewed Hughes's proposed gas prohibition. Their final report concluded that a total ban on chemical warfare, more so than a partial restriction prohibiting its use on civilians, was necessary because "there can be no actual restraint of the use by combatants on this new agency of warfare, if it is to be permitted in any guise. The frightful consequences of the use of toxic gasses if dropped from airplanes on cities stagger the imagination" ("Report on new agencies," 1921).

The Subcommittee on the Limitation of Land Armaments chaired by John J. Pershing, a second working group at the Washington Conference, agreed. Pershing wrote in the subcommittee's final report that "chemical warfare should be abolished among nations, as abhorrent to civilization. It is a cruel, unfair and improper use of science. It is fraught with the gravest danger to noncombatants, and demoralizes the better instincts of humanity" ("Report of the committee," 1921). No doubt Pershing's words carried additional weight considering his service as commander of the American Expeditionary Forces. The delegates unanimously agreed to the language that became Article 5 of the Washington agreement, prohibiting chemical warfare in "conscience and practice":

> The use in war of asphyxiating, poisonous or other gases, and all analogous liquids, materials or devices, having been justly condemned by the general opinion of the civilized world and a prohibition of such use having been declared in treaties to which a majority of the civilized Powers are parties.
>
> The Signatory Powers, to the end that this prohibition shall be universally accepted as a part of international law binding alike the conscience and practice of nations, declare their assent to such prohibition, agree to be bound thereby as between themselves and invite all other civilized nations to adhere thereto. (Boserup, 1973, pp. 153–154)

The United States, Britain, Italy, Japan, and France could not agree on other aspects of the treaty, and it never entered into force, but these negotiations indicate that the principal of prohibiting poison gas weapons enjoyed broad international support three years after the end of World War I. A consensus formed at the conference that the practice of chemical warfare posed a threat to noncombatants. Significantly, the delegates rejected the notion of restricting the use of chemical weapons as a method of protecting civilians and, instead, supported outright prohibition.

This effort continued at Geneva in 1925. On the second day of the Conference on Regulating the International Trade of Arms, Munitions, and Implements of War, the head of the US delegation, Theodore E. Burton, proposed that the conference resume the unfinished work of the Washington Conference by acting to restrict trade in chemical weapons. The idea enjoyed extensive support among the represented governments, and Burton's proposal eventually evolved into a larger discussion on combating chemical weapons traffic by prohibiting chemical warfare. Since this exceeded the original purpose of the conference, the chemical warfare prohibition was moved to an ancillary agreement known as the Geneva Gas Protocol.

With assistance and direction from the Department of State, the US delegation crafted proposed language for the Geneva Gas Protocol based on Article 5 of the draft Washington Conference agreement. The delegation from Poland proposed additional language that prohibited bacteriological warfare, which also had universal support. The Geneva Gas Protocol was signed by the United States and other delegations on June 17, 1925, prohibiting poison gas use "as a part of International Law, binding alike the conscience and the practice of nations" (Boserup, 1973, p. 155). The Geneva Gas Protocol was transmitted back to the United States for Senate ratification, but there it faced opposition.

The Protocol was opposed by the US Army Chemical Warfare Service, who felt that any chemical warfare prohibition put national security at risk. They believed the Protocol would be violated in the future, as The Hague Declaration was during World War I. Amos A. Fries, chief of the Chemical Warfare Service, in an August 1925 memorandum to the secretary of war, urged a strong stand against Senate ratification, writing that "<u>any</u> international agreement covering a prohibition of the use of chemical warfare is fundamentally unsound" because "it cannot be made effective" (Fries, 1925, p. 4).

Earl J. Atkisson, US technical advisor on chemical warfare at the Geneva Conference and former commanding officer of the American Expeditionary Forces 1st Gas Regiment, predicted that public opposition to chemical weapons would be temporary and argued that chemical warfare work should be allowed to continue. "War is abhorrent to the individual," Atkisson wrote in his report from Geneva, "yet he accepts blowing men to pieces with high explosive, mowing men down with machine guns, and even sinking a battleship in mid-ocean with its thousand or fifteen hundred men being carried to certain death" (Atkisson, n.d., p. 4). He and other chemical weapons proponents believed that poison gas would one day become similarly acceptable. The War Department was convinced to adopt this position, and outside organizations—including professional chemistry associations, national defense preparedness groups, and chemical businesses—likewise lobbied against the Gas Protocol.

Other prominent outside figures and policy makers urged ratification when the Senate took up the Geneva Gas Protocol for consideration in December 1926. John J. Pershing wrote a strongly worded letter to the Senate arguing that the Protocol was a necessary security measure to prevent chemical warfare and protect vulnerable civilians. "To sanction the use of gas in any form," Pershing wrote, "would be to open the way for the use of the most deadly gasses and the possible poisoning of whole populations of noncombatant men, women and children" (Pershing, 1926). US Representative Hamilton Fish III, a veteran officer of the 369th Infantry Regiment, sent a passionate appeal to the Senate urging them to ratify the Gas Protocol and "strike a powerful blow at this new weapon of destruction before it becomes the abomination and desolation of modern civilization" (69 Cong. Rec., 1926, p. 368).

Hopes for speedy passage of the Geneva Protocol were dashed, however, in part by the same isolationist sentiment that prevented the United States from joining the League of Nations. Some in the Senate represented constituencies that wanted the United States to remain unfettered by obligations imposed by international agreements. Strong resistance by Protocol opponents, such as Senators James W. Wadsworth and David A. Reed, convinced treaty supporters in the Senate to withdraw it without a vote four days after it was introduced. Wadsworth wrote of the Gas Protocol, "I hope we have given it an eye black enough to dissuade the next Congress from having anything to do with it" (Wadsworth, 1926). It was an understatement; the Protocol was not considered again by the US Senate for almost 50 years.

Despite the Senate's decision, the principal of restricting chemical warfare had clearly taken root in the United States. The US Army Chemical Warfare Service and its projects struggled with anemic support from the War Department through the 1920s and 30s (Faith, 2014). A US delegation again tried to negotiate a poison gas prohibition at the failed World Disarmament Conference in Geneva in 1932. After the United States was drawn into World War II by the attack on Pearl Harbor, the country refrained from using chemical weapons in combat operations. Opposition to chemical warfare was formally affirmed as an element of US security policy on June 8, 1943, when President Franklin D. Roosevelt issued a statement warning the

Axis powers against poison gas use. In his statement, Roosevelt asserted that the United States would not use poison gas weapons except in retaliation. "Use of such weapons," Roosevelt said, "has been outlawed by the general opinion of civilized mankind. This country has not used them, and I hope that we never will be compelled to use them. I state categorically that we shall under no circumstances resort to the use of such weapons unless they are first used by our enemies" (Rosenman, 1950, p. 243).

In 1969, President Richard Nixon reaffirmed US policy against chemical warfare and pledged the destruction of chemical weapons stockpiles in response to volatile public opinion directed at US actions during the Vietnam War. The Senate reconsidered the Geneva Gas Protocol and ratified it unanimously in 1974. Afterward, the United States began negotiations with the Soviet Union and other nations that in 1993 resulted in the Chemical Weapons Convention, a comprehensive international arms control agreement with extensive provisions for verification.

The failure of The Hague Declaration to restrain the use of poison gas projectiles and the widespread use of chemical weapons in World War I did not blunt the poison gas taboo that predated the conflict. Instead, the United States and other nations that emerged from World War I worked to craft chemical warfare agreements with renewed necessity. Postwar agreements demonstrated an increasing appreciation for some of the complexities of enforceable arms control, and this understanding continued to evolve over time. Postwar chemical warfare prohibitions were often justified by a concern for the ways in which poison weapons threatened noncombatants, the same concern expressed by opponents of later weapon systems considered "weapons of mass destruction."

References

69 *Congressional Record* (1926, December 13).

Atkisson E. J. (n.d.). Report on chemical warfare in connection with the conference for the control of the international trade in arms, munitions, and implements of war. Secret and Confidential Files, Records of the Office of the Chief, Record Group 175 (Entry 4, Box 9, Folder 319.1-1997), National Archives, College Park, MD.

Boserup, Anders (1973). *The problem of chemical and biological warfare, volume 3, CBW and the law of war*. Stockholm: Almqvist & Wiksell.

Davis, Calvin DeArmond (1962). *The United States and the first Hague peace conference*. Ithaca: Cornell University Press.

Faith, Thomas I. (2012). "We are still letting that building alone:" The mustard gas plant at Edgewood Arsenal, Maryland, in World War I. *Journal of America's Military Past*, 37(3), pp. 29-42.

Faith, Thomas I. (2014). *Behind the gas mask: The U.S. Chemical Warfare Service in war and peace*. Urbana: University of Illinois Press.

French condemns the use of gas (1915, July 12). *The New York Times*, p. 3.

Foulkes, C.H. (1934). *"Gas!" The story of the Special Brigade*. Edinburgh and London: William Blackwood & Sons Ltd.

Fox, Albert W. (1918, August 13). Oil smudge shows after patrol drops depth bombs on hun. *Washington Post*, pp. 1, 2.

Fries, Amos A. (1921, October 17). [Memorandum to the Assistant Chief of Staff, War Plans Division]. Secret and Confidential Files, Records of the Office of the Chief, Record Group 175 (Entry 4, Box 1, Folder 1), National Archives, College Park, MD.

Fries, Amos A. (1925, August 20). [Memorandum to the Secretary of War]. Charles H. Herty Papers, 1860-1938, Stuart A. Rose Manuscript, Archives, and Rare Book Library (Series 6.3, Box 90, Folder 11), Emory

University, Atlanta, GA.

Fries, Amos A., Clarence J. West (1921). *Chemical warfare*. New York: McGraw Hill.

Gempel, E.P.H. (1925, October 1). The protection of the coast defenses of Manila & Subic Bay against gas. Secret and Confidential Files, Records of the Office of the Chief, Record Group 175 (Entry 4, Box 15, Folder 354.2/1-8), National Archives, College Park, MD.

General Orders, Adjutant General's Office, for 1863 (1864). Washington, GPO.

Herty, Charles H. (1921, October 25). [Letter to S. Whetmore]. Charles H. Herty Papers, 1860-1938, Stuart A. Rose Manuscript, Archives, and Rare Book Library (Series 6.3, Box 77, Folder 7), Emory University, Atlanta, GA.

Higgins, A. Pearce (1904). The Hague conference and other international conferences concerning the laws and usages of war: Texts of conventions with notes. London: Stevens and Sons, Limited.

Lefebure, Victor (1921, January 5). Chemical disarmament. *Chemical and Metallurgical Engineering*, 24(1), pp. 5-7. Charles Evans Hughes Papers, Manuscript Division (Box 160, Folder: Washington Conference), Library of Congress, Washington, DC.

Lighthouse crew gassed (1918, August 13). *The New York Times*, pp. 1, 4.

Mayor, Adrienne (2003). *Greek fire, poison arrows & scorpion bombs: Biological and chemical warfare in the ancient world*. Woodstock, NY: Overlook Press.

Memorandum (n.d.). John J. Pershing Papers, Manuscript Division (Box 81, Folder: Gas, Poison), Library of Congress, Washington, DC.

Miles, Wyndham D. (1970). The idea of chemical warfare in modern times. *Journal of the History of Ideas*, 31(2), pp. 297-304.

N.B. Scott "gassed" (1918, August 4). *Washington Post*, pp. 1, 8.

Pershing, John J. (1926, December 10). [Letter to William E. Borah]. John J. Pershing Papers, Manuscript Division (Box 81, Folder: Gas, Poison), Library of Congress, Washington, DC.

Price, Richard M. (1997). *The chemical weapons taboo*. Ithaca: Cornell University Press.

Report of the committee on limitation of land armaments (1921, November 30). John J. Pershing Papers, Manuscript Division (Box 81, Folder: Gas, Poison), Library of Congress, Washington, DC.

Report on new agencies of warfare (1921, December 1). Special Reports Prepared by the Advisory Committee 1921-1922, International Conference Records, Conference on the Limitation of Armaments, U.S. Delegation, Record Group 43 (Entry 94, Box 1, Folder: Report on Poison Gas), National Archives, College Park, MD.

Roberts, A.A. (1915). *The poison war*. London: William Heinemann.

Rosenman, Samuel I. (1950). *The public papers and addresses of Franklin D. Roosevelt: 1943, the tide turns*. New York: Harper.

Scott, James Brown (1920). *The proceedings of The Hague peace conferences: Translation of the official texts*. New York: Oxford University Press.

Spiers, Edward M. (1986). *Chemical warfare*. Urbana: University of Illinois Press.

Steen, Kathryn (2014). *The American synthetic organic chemicals industry: War and politics, 1910-1930*. Chapel Hill: University of North Carolina Press.

Thomas, Andy (1985). *Effects of chemical warfare: A selective review and bibliography of British state papers*. London: Taylor & Francis.

Tuorinsky, Shirley D. (2008). *Medical aspects of chemical warfare*. Washington: Office of the Surgeon General.

U-boat attempts to gas U.S. city (1918, August 13). *Chicago Daily Tribune*, pp. 1, 4.

Wadsworth, James W. (1926, December 17). [Letter to Charles H. Herty]. Charles H. Herty Papers, 1860-1938, Stuart A. Rose Manuscript, Archives, and Rare Book Library (Series 6.3, Box 90, Folder 11), Emory University, Atlanta, GA.

9

The AEF in the Trenches: The American Military and Modern Warfare in the First World War

Jonathan A. Beall
University of North Georgia

Abstract

The World War strained the abilities of the US Army severely in the 18 months that America was in the war. The US Army had begun to prepare for overseas conflict as a result of the earlier Root Reforms but it took the World War to show how far the US Army had to go to effectively wage modern war. This chapter demonstrates how the operational and tactical conduct of American Expeditionary Force (AEF) were heavily influenced by political and strategic decisions. This chapter is a reminder that the four levels of war are interactive and affect each other, which should complicate how we assess and examine a wartime army's conduct and effectiveness.

Keywords: Policy, politics, strategy, operations, tactics, Elihu Root, American Expeditionary Force, John J. Pershing, Woodrow Wilson, amalgamation, US Army, National Guard, St. Mihiel, Meuse-Argonne, Lorraine, Operation

Michael, France, Great Britain, 1st Division, 2nd Division, 3rd Division, 35th Division, effectiveness

The AEF in the Trenches: The American Military and Modern Warfare in the First World War

World War I presented the US Army with a challenge that was unprecedented in size, scope, and scale. In many ways, not even the Civil War in the 1860s could prepare the army for what was demanded of it after the United States declared war on Germany in April 1917. As such, World War I became an incredible test of the US Army's ability to mobilize, plan, and combat a veteran enemy waging industrialized war. Although the United States came up short in many ways on the battlefields of 1917 and 1918, the army spent much of the 1920s and 1930s mulling over its World War experiences and institutionalizing those conclusions for future conflicts, applying many of these lessons in World War II.

The United States entered the war late and from a standing start with nearly no preparations logistically or strategically. The infantry doctrine was ill-suited for the battlefield in 1917 and 1918, and there was little operational thinking in the US Army at this time. The American military, from the senior political and military leadership in Washington, DC, to General Pershing's headquarters in France to the doughboys in the trenches, was going to have to adapt and learn as the military prepared to fight the Germans. For those fighting, the buoyant optimism that led most Americans to think they could whip the Germans died an ugly death at places like Belleau Wood, on the Marne River, or at the Meuse-Argonne. Many of the lower-level decisions that affected American operations and tactics, however, were directly influenced by the larger political and strategic decisions. America's wartime experience in 1917 and 1918 is a reminder that the levels of war are not discrete, disparate levels but overlapping concepts that continually influence each other. The experience of the American Expeditionary Force (AEF) in World War I shows how the four levels of war—politics, strategy, operations, and tactics—interact and affect each other, which complicates a nation's war effort and assessment as well. This essay examines where the US military stood when the country declared war on Germany in 1917 and demonstrates how the political and strategic levels of war affected American war effort at operational and tactical levels, especially as America was forced to adapt to radically different methods of warfighting than it initially planned. Ultimately, a study of the AEF in the trenches must show the complexity of a nation's war effort and demonstrates how the levels of war are not discrete levels but interacting elements that affect and influence each other.

US Army, 1903–1917

By 1917, the US Army had experienced significant structural changes since the preceding war against Spain in 1898 and certainly since its last major war, the Civil War, but it was still influenced by its recent past. By April 1865, there were over one million soldiers in Union uniform, but, pursuant to its citizen-soldier tradition, the American government had demobilized up to 57,000 within a year, and the army floated around the 25,000-man mark for much of the postwar period. It was not until the Spanish-American War that the size of the army increased (Weigley, 1967, p. 567-8). West Point, however, failed to teach anything beyond minutiae to the cadets in the decades afterward. Dwight D. Eisenhower recalled that, while a cadet, he memorized troop locations at the Battle of Gettysburg but not its larger strategic significance (Seidule, 2014, p. 8).

The American frontier was a prevailing influence upon the character, nature, and organization of the US Army, which minimized preparation for modern, industrialized war. The frontier had been the army's primary concern since the earliest days of the republic. Although historians date the closing of the frontier and the end of the Indian Wars in 1890, the army continued to man forts throughout the American West. Indeed, by early 1917, a sizable portion of the army was in northern Mexico chasing after Mexican bandit Pancho Villa. For decades, manning these frontier posts forced units to decentralize, so, worsened by low budgets, large-unit training rarely occurred through much of this period. For example, when army leadership tried to mobilize a 13,000-man division for the Texas–Mexico border, it took 3 months to bring the men together (Weigley, 1967, pp. 334-335).

After 1898, the army received the mission of defending the nation's empire, including the Philippines and Hawaii. Historian Brian Linn (1996) asserts that this "Pacific frontier duty" likely affected more officers than the American frontier, and it certainly extended the frontier mindset into the 20th century (pp. 142, 167). Not least among its influence, this frontier constabulary role helped to ensure an insular, inward perspective of continental defense for American military planners. As such, the lieutenants and captains who cut their teeth at frontier posts in the American West or in the Philippines, leading small units with minimal large-unit training in a comparatively tiny army, were the senior generals in 1917.

While frontier defense had an important influence upon the US Army into the 20th century, there were important steps that moved the army away from that insular perspective toward being able to enter global conflicts against opponents more heavily armed than the Sioux Indians or Filipino *insurrectos*. Between the Civil War and World War I, there were several officers who considered how the army might plan for, mobilize, and fight in a major European conflict. One of the more important thinkers was Emory Upton, a West Point graduate who commanded a brigade of volunteer citizen-soldiers in the Civil War. Upton sought to repair a glaring error that the Civil War revealed in American manpower policy: the use of a small cadre of professional Regular Army soldiers and a much larger body of volunteer citizen-soldiers. Upton was bothered by the unprepared, untrained nature of these citizen-soldiers and the extremely high numbers of casualties caused by a lack of prewar preparation. Upton proposed the creation of a voluntary, federally-controlled army reserve that could train American men for war but not necessarily have them serve in the Regular Army. Upton worked to retain America's sacrosanct faith in the citizen-soldier but also prevent the costly inefficiencies that he believed would otherwise occur in the nation's next wars (Fitzpatrick, 2017, pp. 186-189). Upton's critique of America's blind faith in the untrained citizen-soldier affected American military thinkers into the 20th century.

The short Spanish-American War in 1898 revealed many structural problems in the US Army. In the wake of the scandalous performance of the War Department, President William McKinley replaced his overwhelmed Secretary of War Russell Alger with New York lawyer Elihu Root. Root worked with a number of reform-minded officers to modernize the army and prepare it for a large overseas conflict. Because of the Root Reforms, the federal government (and, by extension, the US Army) more directly controlled the National Guard, although not completely. Root also created the Army General Staff, headed by the chief of staff, to advise the president and secretary of war, centralize army administration, and make plans for possible future wars. The last significant reform created the Army War College, which housed the general staff as well as its planning contingents and trained senior officers to think strategically. Upon these improved strategic and educational bases, Root and these army reformers hoped to create an army capable of waging modern war overseas. From Root's mindset, the purpose of the army was not continental defense; instead, he argued, "the real object of having an army is to provide for war" (United States War Department, 1899, p. 45). World War I tested the effectiveness of these reforms.

American infantry tactics rested on the ideas of "open warfare," which relied almost exclusively upon the infantryman with the bayonet-tipped rifle. It relied upon well-trained soldiers, accurate with their rifles, to maneuver against the enemy in order to achieve tactical dominance on the battlefield and win operational victories. Although the machine gun and American artillery developed during this period, "open warfare" had little use for either automatic weapons or infantry-artillery training because army doctrine perceived the infantryman and his bolt-action rifle as the decisive elements on the battlefield.[1] This system was human-centered rather than based upon machines and equipment. Beginning with senior leaders, open warfare influenced American military assumptions on the tactical and operational levels.

While these reforms were important steps, the army's older cultures and systems simply made room for new agencies, new ways, and new officials without really being transformed. The Root Reforms were essential for the modernization of the US Army, but they did not change the army immediately, and they were not as transformative as hoped. While army officers were cognizant of changes in war during these decades and were attentive to the Great War itself, the US Army retained strong connections to its 19th-century constabulary ways in April 1917, if only because its senior leaders were accustomed to that mindset. For example, the Regular Army generally remained biased against the National Guard and its officers during this period and through World War I despite attempts to tie the two organizations closer together. Historian Richard Faulkner (2012)

---

1 The best description of open warfare is Grotelueschen (2007), Faulkner (2012), and Rainey (1983).

asserts that, in 1917, the US Army operated and fought like an army from 1914: the US entered World War I in 1917 with a similar mindset and perspective as the other nations had in 1914, regardless of the intervening 3 years of carnage and change (Faulkner, 2012, p. 327; Clark, 2017, pp. 225, 258-260).

Key Political and Strategic Decisions

Understanding the mindset and culture of the US Army in 1917 establishes an important context to explain what the army could and could not do when it entered the war. At the same time, it is also important to consider how early wartime decisions also affected the American Expeditionary Force. Four decisions, made at the political and strategic levels of war, directly influenced the operational and tactical levels of the American war effort in 1917 and 1918.

The first crucial decision was President Woodrow Wilson's political objective that guided the American war effort and influenced nearly everything else. When the United States declared war in April 1917, Wilson made important early decisions—based upon his political objectives—that affected the formation, contribution, and effectiveness of the AEF. To begin, when the United States declared war on Germany, it was not readily apparent what America's contribution would be: financial, material, or personnel. One of Wilson's chief political goals was to be able to influence the peace treaty at the end of the war. His clarification of the Fourteen Points in January 1918 helps demonstrate the political influence that he hoped to have on the postwar order. Simply providing money and material would not give Wilson the leverage he needed to accomplish that goal: America's contribution would have to be a strong military one. The US Army concurred with that assessment as well as with the conclusion that the

Western Front had become the decisive front by 1917 (Millett, 1986a, p. 235; Nenninger, 2010, p. 126).

That first decision—America's political objective—influenced the second major decision, which turned into a persistent debate between the United States and its partners over the nature of that American military contribution. American leaders determined that an American army had to contribute significantly to an Entente victory in France, which meant an independent American army (Millett, 1986a, p. 235; Nenninger, 2010, p. 126). This decision to create a discrete American Expeditionary Force ran counter to urgent, persistent French and British requests to amalgamate, or incorporate, American manpower into their veteran, albeit exhausted, armies. After the losses and failures in 1917, the French and British were concerned they would not be able to defeat a German military reinforced by its forces from the east after Russia left the war (Bruce, 2003, pp. 144-145). There the American soldiers would be trained in already-existing training facilities and be supplied by already-existing logistical systems. The Entente argued that amalgamation was better than waiting for the United States to gather, build, train, and ship a complete army overseas. The Americans, pursuant to their political objectives and influenced by national pride, remained steadfast in their goal to create an independent American army that maintained its own section of the Western Front. The amalgamation controversy strained the already-stressed Entente coalition, and it persisted into the first half of 1918, but the Americans held firm to the policy of creating an independent American army to fight alongside the French and British, even as a few American divisions did train with the French and British on the front lines (Trask, 1993, pp. 5-6, 12, 39; Bruce, 2003, pp. 144-170; Stevenson, 2011, pp. 44, 247-249).

These two key decisions had a strategic effect upon the American contribution. To influence the postwar peace, there had to be an independent

American force that made significant and decisive contributions to defeating the Germans. That, therefore, required a mass, national army to fight alongside the French and British. Knowing that the 300,000 men in the US Army and National Guard were insufficient, the United States used conscription to build a mass national army. To that end, the AEF commander, General John Pershing, received explicit orders that "the forces of the United States are a separate and distinct component of the combined forces, the identity of which must be preserved" (quoted in Trask, 1993, p. 12). Because American planners also knew this "separate and distinct" army would take time to build, strategists planned for capable, trained American forces to begin launching major offensives against the Germans by 1919 at the earliest (Coffman, 1986, p. 18; Trask, 1993, p. 6). In July 1917, Pershing predicted the AEF would be around three million men by 1919 (Coffman, 1986, p. 127).

The decision to build a mass army made political and strategic sense, but actually achieving that goal was something else entirely. One major problem that the Americans faced was that of industrial mobilization and supply. After declaring war in April, the US government quickly passed the Selective Service Act in May in order to more efficiently create that independent American army. Very rapidly, the US Army grew from over 100,000 men to over a million. By war's end, more than three million American men had served in uniform.

The War Department, still operating through long-standing systems and procedures dating back to the 1800s, was wholly unprepared to arm, clothe, equip, and sustain this massive force. Much of the private arms manufacturing was producing material for the British and French armies and could not easily mass produce American weapons. Furthermore, it would take time for peacetime industries to convert to wartime production. In short, this massive independent American force lacked the sufficient numbers of machine guns, small arms, artillery, airplanes, and tanks. When it became clear how long it would take to convert peacetime industrial capabilities to wartime production, the French and British agreed to arm the Americans. American soldiers fired British rifles (rechambered for American bullets) and machine guns while they relied heavily upon the French for tanks, airplanes, artillery, and automatic rifles. The industrial situation in the United States became snarled and chaotic through 1917, but it was becoming more efficient and effective by the end of the war in 1918. Still, by November 1918, there were few American-made machine guns, artillery, tanks, or airplanes on the front lines (Bruce, 2003, pp. 100-105).

The third major decision that affected the AEF on the front line was Pershing's military strategy that included where and how the United States' forces would fight on the Western Front. Looking at a map, Pershing decided that the AEF's military contribution would be in France's Lorraine region. This decision was affected as much by logistics as anything else. By placing Americans in Lorraine, they could rely upon ports of entry as well as railway lines from St. Nazaire and La Pallice on the Atlantic as well as Marseille on the Mediterranean, and not strain French and British lines of communication (Vandiver, 1977, pp. 730-731). In Lorraine, once properly trained and organized, the AEF planned to launch a major offensive against Metz in 1919. The strategic goal was to break through the German lines and seize the industry and railways east of Metz along the Saar River. Pershing expected this strategy to allow America to make a decisive contribution to Germany's defeat. This plan called for a preliminary offensive to flatten the bulge into French lines—the St. Mihiel salient—prior to the major offensive to capture Metz and the Saarland beyond (Coffman, 1986, pp. 125-126; Millett, 1986a, pp. 238-239). Whether achieving these strategic objectives would have been as decisive as expected is debatable, but it helps to explain why the AEF fought where it fought in World War I as well as its strategic mindset as it entered France.

Having decided to create an independent army fighting on its own sector of the Western Front, the United States had to build and train that force. A key implication of these decisions was the army's expectation to fight with open warfare, which directly affected the operational and tactical abilities of the AEF through 1917 and 1918. Commanding general of the AEF, John Pershing, oversaw a dizzying array of complexities, issues, challenges, and questions as he built his independent American force. One issue he had to consider was the infantry doctrine that the Americans would use to guide their fighting against the Germans. Pershing was convinced that America's open warfare would provide the combat strength and mobility necessary to pierce the German lines and regain momentum to the modern battlefield. As stated earlier, by waging open warfare, the Americans hoped to restore movement and maneuver to the battlefield through the American rifleman and his bayonet-fixed rifle. Pershing accepted the assumptions of open warfare and argued that stateside training must emphasize that "the rifle and the bayonet remain the supreme weapons of the Infantry soldier" because "the ultimate success of the army depends upon their proper use in open warfare" (Grotelueschen, 2007, p. 31).

Pershing was appalled at the tactics of trench warfare that had developed and that had influenced French and British infantry by 1917, and he was determined not to adopt them. The tactics of trench warfare, as they had developed by 1917, emphasized firepower and machines rather than humans. They included massive artillery bombardments; heavy reliance on machine guns; small-unit tactics that incorporated various weapons such as automatic rifles, trench mortars, and grenadiers; increased coordination between infantry and artillery; and controlled offensives that sought to achieve limited gains rather than decisive breakthroughs. As the war dragged on, the Allies' tactics also incorporated tanks and close air support from purpose-built fighter planes. Insofar as trench warfare tactics had evolved by emphasizing technology over man, World War I combat fundamentally challenged, if not overturned, the very nature of open warfare (Grotelueschen, 2007, pp. 30-35).

Leveraging American power on the battlefield included organizing American divisions into two brigades of two infantry regiments each—and included separate machine gun battalions as well as field artillery regiments—which made the American division nearly twice the size of European divisions (28,000 compared to approximately 14,000 in the European divisions). They were also intended to generate more firepower as well as absorb higher losses (Trask, 1993, pp. 18-19).

As such, the decision to build an independent American army that could wage open warfare affected conscripted officers' and soldiers' training in the United States. To be sure, draftees' training was complicated by the utter unpreparedness of the War Department: after entering training facilities constructed from scratch, many American conscripts trained with old weapons or simply made wooden cutouts. While plans were made with America's coalition partners to help arm the AEF, the mass of draftees had to make do with clothing, equipment, and weapons shortages that hampered training. Many soldiers only received key weapons and pieces of equipment when they reached France (Ferrell, 2007, pp. 15-16; Faulkner, 2012, pp. 108-109).

Just as the army struggled to supply its men with needed weapons for training, so also it struggled to train its men for trench warfare. Pursuant to Pershing's embrace of open warfare, the plan was to train soldiers to become proficient in those skills. Training for thousands of newly commissioned officers in the expanded national army heavily emphasized close order drill, musketry and marksmanship, and bayonet drill, but did not include nearly as much trench warfare tactics nor infantry–artillery cooperation (Faulkner, 2012, pp. 35-46). Pershing intended for the War Department to train the AEF divisions stateside in the methods of open warfare and then undergo a three-phase training program lasting 3 months once the various divisions reached France. Because of Pershing's faith in open warfare, the general intended the independent American army to become proficient in open warfare but within the context of trench warfare (Rainey, 1992, p. 95; Faulkner, 2012, pp. 141-144). To be clear, many American divisions received some instruction and training in trench warfare from the British and the French, which were at odds with the ideals of open warfare (Faulkner, 2012, pp. 126-128, 144, 147-148; Bruce, 2003, pp. 120, 143; Trask, 1993, pp. 18-20). There was confusion among soldiers and junior officers trying to reconcile the two competing tactical doctrines, but those divisions that most embraced the thinking and mindset of trench warfare typically performed better in combat. Because American observers had seen earlier that open warfare would not work as well as expected, historian Richard Faulkner (2012) is highly critical of the US Army's failure to grasp that the different nature of World War I combat had rendered open warfare tactics ineffective; consequently, the army generally trained American citizen-soldiers and officers ineffectively and inadequately for the realities of the war in France. Further, Faulkner argues American military leadership also systemically failed to detect the need to learn and adapt to new operational and tactical circumstances (pp. 9, 46-47, 316-317).

These three key decisions were made during 1917, but the last one came in early 1918. In part to win the war before the Americans made themselves felt on the battlefield, the German army launched major attacks against the British and French in the spring of 1918, beginning with Operation Michael. Germany's Operation Michael rousted the British and French armies out of their trenches and threatened Paris, but it ran out of steam by June. In those desperate weeks, getting American manpower across the Atlantic became a top priority and the fourth major strategic decision. After the Germans launched Michael in March, British Prime Minister David Lloyd George urged President Woodrow Wilson to get 120,000 men *a month* into France. However, there was a heavy priority on infantry and machine gunners rather than on entire divisions replete with their combat support and logistical infrastructure. Several American units were temporarily placed into, but not amalgamated with, British and French units, which, at the very least, allowed these Americans to learn from veterans. Twelve American divisions deployed to the British front for training and emergency usage (Millett, 1986a, p. 243). This decision dramatically increased the American presence in France. Of the 2.8 million draftees who served in World War I, 2.3 million entered service and arrived in France in 1918 (Stevenson, 2011, pp. 249-250; Millett, 1986a, p. 242).

These German attacks were a reminder that "the enemy gets a vote" as they thoroughly upended many of Pershing's strategic plans and objectives, especially the plan to build a trained army and launch major offensives in 1919. One consequence of this decision to rapidly send American forces to France because of the German offensive: many American divisions went partially trained, if at all. The offensive and the decision to move nearly two million men overseas by November 1918 meant that training was severely curtailed stateside and was nearly nonexistent for those in Europe. Through 1918, divisions arrived in Europe having completed various stages of training in the United States and finished abbreviated training camps in France, thanks to the changed strategic picture. By mid-1918, very few American divisions had completed the full 3-month training cycle in France. These circumstances meant that Pershing flung ill-trained Americans against veteran German units in the Meuse-Argonne offensive from September to November 1918 (Neninger, 2010, p. 149).

Another result of this decision affected the American supply system. Packing combat personnel into the holds of ships meant fewer trucks and draft animals to support and supply those combat soldiers. The lack of supply infrastructure directly affected the ability to sustain combat forces in the field. It proved difficult to organize supply lines, feed soldiers, retrieve and evacuate the wounded, and bury the dead as major traffic jams in the rear areas hindered mobility and communications (Ferrell, 2007, p. 5; Nenninger, 1987, pp. 180-181; Lengel, 2008, p. 144; Faulkner, 2019, 489). Supply of frontline forces sometimes became ad hoc and improvised rather than systematic and predictable, and American doughboys simply took what they wanted from unsecured and unmonitored dumps (Nenninger, 1987, p. 180). Because of the shaky supply situation and inexperienced junior officers, historian Edward Lengel (2015) writes that American soldiers "were practically reduced to the status of beggars" compared to the French and British who enjoyed well-established supply systems (p. 366). Though the American forces learned from their World War I experiences, those lessons came at a steep price.

The AEF in the Trenches

These four decisions—Wilson's political objectives, America's decision to create an independent national army, the decision to fight in Lorraine, and the decision to rush partially trained divisions to France—were made at the political and strategic levels. These decisions directly influenced American effectiveness at the operational and tactical levels. Ill-trained American soldiers fought as an American army but with British and French weapons while oftentimes employing infantry doctrine and tactics that the World War I battlefield bloodily and consistently showed to be inadequate, supported by a supply system that teetered on the brink of collapse. In sum, the American war effort on the Western Front can be seen in three phases: (1) the limited actions of the initial five divisions in helping to stem the German 1918 offensive, (2) the actions in the Allied counteroffensive of 1918, and (3) America's two major offensives at the St. Mihiel salient and the Meuse-Argonne.

Amidst great fanfare, the United States formed its 1st Division and sent it to France in 1917. However, it was filled with new recruits and draftees that required many months of training under the watchful eyes of French officers. Ultimately, only the 1st, 2nd, 26th, and 42nd Divisions enjoyed a strong training regimen in trench warfare under veteran French influence. It is no coincidence that the 1st and 2nd Divisions became two of the AEF's best divisions. The British also helped to train the 27th, 28th, 30th, and 82nd Divisions (Faulkner, 2012, pp. 151, 272).

As the first American divisions trained in France and became accustomed to life in the trenches through 1917 and into 1918, there were minor raids and skirmishes against the Germans that taught the Americans the bloody nature of their business. In March 1918, the German-launched Operation Michael struck against the French and British lines further north and did not directly affect the Americans. During this time, as mentioned earlier, the Americans accelerated the pace of shipping personnel, and, as a result, its in-theater training cycle virtually collapsed.

During this period, the Americans experienced their first major engagements by helping to stem the German offensive. In the 1st Division's fight for Cantigny on May 28, 1918, the division employed trench warfare techniques—given their months of training with the French—and successfully seized the town from the Germans. In his description of the attack on Cantigny, historian Mark Ethan Grotelueschen (2007) argues that the division did away with open warfare thinking and embraced the ideas of trench warfare at Cantigny (Grotelueschen, 2007, pp. 74-83; Millett,

1986b, pp. 179-182). The 1st Division is Grotelueschen's best example of a World War I American division putting aside the traditional American concept of open warfare and embracing the more relevant tactics of trench warfare. While the division faltered at Soissons at the Aisne-Marne offensive, the 1st Division applied the hard-learned lessons of trench warfare taught by its French army tutors and generally fought well in World War I as one of America's elite divisions.

Grotelueschen (2007) uses the 2nd Division to demonstrate the process of a division learning in wartime. The 2nd Division was unique because one of its two brigades were the 5th and 6th Marine Regiments, making it a half-army, half-marine division. The 2nd Division came together in September 1917, but division training did not begin in earnest until January 1918 (pp. 202-203). Though the German offensive cut short this training, much of it had been completed, including time in trenches learning the art of trench warfare. In early June, the Marines of the 2nd Division attacked through Belleau Wood utilizing open warfare tactics and paid heavily for their attempts. They failed to use artillery support and suffered as a result, but when they began to use artillery bombardments before an attack, seizing Belleau Wood became much easier. The attack on Belleau Wood began on June 6, and the area was declared clear of Germans on June 26.

In his study of the marines in Belleau Wood, Lengel (2015) indicts 2nd Division commander General James G. Harbord with poor leadership. While Lengel argues that the two regimental commanders made many mistakes, he reserves his strongest critique for General Harbord. Harbord, according to Lengel, was unwilling to deviate from open warfare tactics, failed to understand the difficult conditions on the ground, and relentlessly pushed the marine regiments to take the forest despite the heavy German opposition. Harbord's dislike of French "caution" was a result, Lengel asserts, of Harbord's failure to appreciate French tactical methods by 1918 (pp.

129-135, 200, 205-206, 300). At the same time, Lengel demonstrates, the 3rd Division fought alongside the French 10th Colonial Division, learned the tactics of trench warfare, and more effectively seized ground from the Germans but at a lower cost in personnel (2015, pp. 128-130, 132, 135).

On July 1, the 2nd Division's army brigade attacked the Germans near Vaux. There the brigade coordinated between infantry and artillery; better combined machine guns, mortars, and automatic rifles with the riflemen; and established smaller, more limited objectives. Additionally, soldiers had learned to suppress German machine gun strongpoints with American machine guns while soldiers outflanked the Germans rather than making frontal assaults. While Belleau Wood showed the bloody results of applying open warfare tactics to the battlefield, Vaux showed the division moving away from that mindset and achieving more success (Grotelueschen, 2007, pp. 206-226).

The savage fighting at Cantigny, Belleau Wood, and Vaux were conducted by divisions that were learning and integrating the lessons of trench warfare into their planning and fighting, albeit unevenly. They also were fought by divisions that had spent more time in France learning these methods than had other divisions. The Meuse-Argonne offensive from September to November 1918 showed that not all American divisions had done likewise, especially those that did not complete the training cycle but were thrust into combat as a result of the crisis caused by the German spring offensive. For example, many accounts across the AEF indicate that officers understood slowly—if at all—that for any attack to succeed, officers had to know how to combine and incorporate a variety of weapons, such as machine guns, mortars, and grenades to help riflemen close with their enemy. Infantry officers also had to know how to coordinate with the artillery, a crucial job complicated by the limitations in communications technology at that time. Much of this failure resulted from AEF officers' training that did not adequately prepare junior leaders on combined-arms tactics and fighting (Faulkner, 2012, p. 256). In general, considering operational and tactical success in some divisions, there are many examples of officers who failed to understand that the World War I battlefield was a far more complex place than open warfare tactics assumed. The end result was American units making frontal assaults against well-sited German defenses with mutually supporting machine gun strongpoints. Faulkner (2012), in his critical view of the AEF, asserts that these failures generally stemmed from AEF junior officers' poor training that utterly failed to give officer-candidates an accurate understanding of World War I combat (p. 272).

The larger offensives at the St. Mihiel salient in mid-September and the Meuse-Argonne from late September until the Armistice in November used a menagerie of effective and less-than-effective American divisions that had varied degrees of experience in trench warfare. To what extent each division applied the ideas of trench warfare rather than open warfare varied from division to division. The 1st and 2nd Divisions, veterans by September 1918, both understood that the rifleman with his bayonet could not use the tactics of open warfare to defeat the Germans. In the Meuse-Argonne, both divisions used artillery preparations and rolling barrages, made limited attacks, and then regrouped to launch other limited attacks. They relied on a mix of weapons to support the riflemen, including machine guns, trench mortars, grenadiers, and automatic rifles. These divisions understood the World War I battlefield and made good progress in their sectors during the Meuse-Argonne.

In contrast, the 35th Division, a National Guard division composed of Missourians and Kansans, became wholly ineffective 4 days into the battle. In his study of the 35th Division and its collapse, historian Robert Ferrell (2004) argues that the division's collapse came from various sources. He cites the main causes as Regular Army officers' prejudice against National Guard units and officers; insufficient training and insufficient artillery support; and a division leadership that made too many command changes but offered poor administration and failed to adequately feed and clothe the men (Ferrell, 2007, p. 72).[2] In short, Ferrell portrays a division that was poorly trained, poorly led, poorly supported, and poorly equipped, finally breaking in front of veteran German forces in the opening days of the campaign. In his study of the Meuse-Argonne, Edward Lengel (2007) portrays a loss of cohesion and leadership at the division and regimental levels. The doughboys in the 35th Division fought hard against the Germans and suffered heavily, but there was organizational collapse by the 29th (Lengel, 2008, pp. 154-155, 172-177, 188). Ferrell observes that this division's failure demonstrates "how tenuous was the victory of the American Expeditionary Forces . . . how difficult were the challenges; how the men of the AEF did the best they could" (Ferrell, 2004, p. x). The process of learning and improving was not general throughout the AEF, and there clearly was a spectrum of effective and ineffective divisions.

At the same time, the supply system during the fighting at the Meuse-Argonne nearly collapsed. Here are the anecdotes of a dysfunctional logistical system that failed these ill-trained American soldiers: as one soldier in the 3rd Division remembered, a ration dump was "just what the name implies—a dump" within a collapsed dugout from which anybody could carry away what they pleased (Nenninger, 1987, p. 180). Lengel (2008) describes French Premier Georges Clemenceau trying to find First US Army headquarters in the first days of the Meuse-Argonne offensive and being trapped in a 5-mile traffic jam. It did not look good for American senior commanders for a head of state to witness such disorder and organizational inertia (pp. 183-184). That uneven process of learning and the uneven operational and tactical effectiveness of American

---

2   Ferrell describes the division's training at Ferrell, 2004, pp. 2-5.

units on the World War I battlefield were influenced by the political and strategic decisions made earlier by President Wilson, his chief advisors, and General Pershing. To be clear, these difficulties were not all American leaders' fault, as the German spring offensive panicked the British, French, and Americans enough to force them to rush the incomplete American divisions to the Western Front.

Assessment

Among historians, there is no consensus on the effectiveness of the AEF. Mark Ethan Grotelueschen (2007) uses the 1st, 2nd, 26th, and 77th Divisions as case studies to point toward a general pattern of learning, adaptation, and effectiveness, although to different degrees (pp. 343-349). Richard S. Faulkner (2012) argues against this interpretation to say that because of the generally insufficient training time and quality, most American divisions failed to learn and adapt adequately. On the battlefield, or the "school of hard knocks," men were killed because of their poor training, only to be replaced by inexperienced noncommissioned officers and commissioned officers who repeated the same mistakes. That "vicious cycle" was the main reason "why the AEF never truly experienced marked improvement in its combat effectiveness during the war" (pp. 286, 293). Robert H. Ferrell (2007), in his analysis of the Meuse-Argonne, makes a similar case to Faulkner's (Ferrell, 2007, pp. 148-155). In his study of the AEF before the St. Mihiel salient and the Meuse-Argonne, historian Edward G. Lengel (2015) argues that the American doughboy's experience was far more varied and complex (pp. 370-374). The AEF demonstrates the complex nature of wartime learning by varying across all military units, from platoons to army groups, and even veteran leaders often made decisions they knew they should not make.

Perhaps even more so than the Civil War in the 1860s, World War I offered a severe challenge to the US Army which had developed within a certain insular, continental context between 1865 and 1917. It was even able to adapt much of its frontier mindset to its small empire. While there were thinkers such as Emory Upton, who challenged this insular view, it remained the predominant mindset until 1917. Within this context, the US Army was quite adept at its given tasks: suppressing indigenous revolts as well as continental, and later imperial, defense. Indeed, the senior leaders of the AEF developed professionally within this inward-looking mindset. Key reforms between 1903 and 1917 helped to move the US Army to an outward-looking mindset capable of fighting major wars overseas, but World War I revealed that it still had a long way to go.

World War I challenged how the United States used military force for large-scale global wars. Because it jumped into the fray mid-war, America—not unexpectedly—failed to develop the political, social, and military institutions that developed in Great Britain, France, Germany, and Austria-Hungary by 1917. That is to say, it is no surprise that America brought many pre-1914 perspectives with it when it entered the war in 1917. It should also not be a surprise that America struggled to project power in 1917 and 1918. Indeed, in his essay on total war in World War I, Roger Chickering (2000) rightly reminds us that 1914 was fought in an older, 19th-century mindset, that it took until 1915 for the warring governments to realize that this war would not be won quickly or cheaply, and that it would become necessary to centralize and commandeer their economies and societies (pp. 37-39, 52-53). America did not enjoy that luxury and had to centralize its government and build the AEF very rapidly.

As the United States prepared to use military force, it faced many challenges. The army was too small by itself to fight America's war, but building a mass, national army through conscription would take time.

Industry could not immediately sustain and support the AEF, even as the Wilson administration created numerous federal agencies to organize the war effort. The army itself was not ready to fight a war so radically different from any in its history. The decisions made in 1917 by political and strategic leaders directly influenced what the AEF did and how well it fought. Wilson's political objective of influencing the postwar peace required a major American contribution. Consistent with that policy objective was the decision to build an independent American army, and the natural inclination to wage open warfare against the Germans affected the entire AEF story. Given that policy objective and strategic objective of creating an independent army, Pershing created a military strategy that would place the Americans in Lorraine capable of large-scale operations in 1919, a plan which took into account the time it would take to build a one- to three-million-man army to help defeat the Germans. The Germans' springtime offensive in March 1918 was a key external force that also affected the AEF's battlefield experiences because the desperate need for manpower forced the last decisions, that is, the early-1918 decision to ship combat personnel without their supply and support staff, which, in turn, influenced how the AEF could sustain its operations. As Grotelueschen (2007) demonstrates, there was learning that occurred in the AEF. However, as Faulkner (2012) and Lengel (2015) show, that learning was neither systemic nor uniform throughout the entire American army.

These decisions affected America's operations and its tactical abilities. The training that most AEF divisions experienced was inconsistent and oftentimes insufficient. Nevertheless, the AEF eludes a general pattern of either effectiveness or ineptitude. Despite their many challenges, some divisions in Europe adapted to the World War I battlefield and fought quite well. The 1st and 2nd Divisions are examples of this adaptation and effectiveness. Many of the partially trained divisions that rushed to the

Western Front in 1918 did not fare so well. These were the AEF units slaughtered as they launched frontal assault after frontal assault into the teeth of chattering German machine guns in the Meuse-Argonne. It is no surprise that the 35th Division, not well led nor well trained, fell apart and was rendered wholly ineffective in the Meuse-Argonne. The operational and tactical abilities and inabilities of the AEF were directly influenced by the political and strategic decisions made by senior American leaders, even if those political and strategic choices were rational and internally consistent or forced upon them by external actors, such as the German offensives in 1918.

In his assessment of American military effectiveness across the four levels of war in World War I, Timothy Nenninger (2010) concludes that "important operational and tactical failings were directly attributable to decisions made at the political and strategic levels" (p. 152). Nenninger likewise observes that these levels are interactive when he writes that "decisions made to improve political and strategic effectiveness, or in pursuit of political and strategic goals, inhibited performance in the operational and tactical realms" (2010, p. 153). As Nenninger observes and as this chapter demonstrates, the AEF's experiences are a reminder that the four levels of war are interactive and they are neither discrete nor disconnected from each other. This chapter demonstrates the varying effects that fundamental political and strategic decisions can have on the operational and tactical levels of war. These experiences show the difficulty in making army-wide conclusions about military effectiveness, especially for a rapidly-created army that fought for only 18 months and whose battlefield contribution was between May and November 1918. They are a reminder that a nation's war effort is a far more complex endeavor than soldiers simply "going over the top."

References

Bruce, Robert B. (2003). *A fraternity of arms: America & France in the Great War*. University Press of Kansas.

Chickering, Roger. (2000). World War I and the theory of total war: Reflections on the British and German cases, 1914-1915. In Roger Chickering and Stig Förster (Eds.), *Great War, total war: Combat and mobilization on the Western Front* (pp. 35-53). Cambridge University Press.

Clark, J. P. (2017). *Preparing for war: The emergence of the modern U.S. Army, 1815-1917*. Harvard University Press.

Coffman, Edward. (1986). *The war to end all wars: The American military experience in World War I*. University of Wisconsin Press.

Faulkner, Richard S. (2012). The *school of hard knocks: Combat leadership in the American Expeditionary Forces*. Texas A&M University Press.

Faulkner, Richard S. (2019). World War I. In Christopher R. Mortenson and Paul J. Springer (Eds.), *Daily life of U.S. soldiers: From the American Revolution to the Iraq War* (Vol. 2, pp. 455-532). ABC-CLIO.

Ferrell, Robert F. (2004). *Collapse at Meuse-Argonne: The failure of the Missouri-Kansas Division*. University of Missouri Press.

Ferrell, Robert F. (2007). *America's deadliest battle: Meuse-Argonne, 1918*. University Press of Kansas.

Fitzpatrick, David. (2017). *Emory Upton: misunderstood reformer*. University of Oklahoma Press.

Grotelueschen, Mark Ethan. (2007). The *AEF way of war: The American Army and combat in World War I*. Cambridge University Press.

Lengel, Edward. (2008). *To conquer hell, the Meuse-Argonne, 1918: The epic battle that ended the First World War*. Henry Holt.

Linn, Brian. (1996). The long twilight of the frontier army. *Western*

Historical Quarterly, 27(2), 141-167. https://doi.org/10.2307/970615

Millett, Allan. (1986a). Over where? The AEF and the American strategy for victory, 1917-1918. In Kenneth J. Hagan and William R. Roberts (Eds.), *Against all enemies: Interpretations of American military history from colonial times to the present* (pp. 235-256). Greenwood Press.

Millett, Allan. (1986b). Cantigny, 28-31 May 1918. In Charles E. Heller and William A. Stofft (Eds.), *America's first battles, 1776-1965* (pp. 149-185). University Press of Kansas.

Nenninger, Timothy K. (1987). Tactical dysfunction in the AEF, 1917-1918. *Military Affairs 51*(4), 177-181. https://doi.org/10.2307/1987946

Nenninger, Timothy. (2010). American military effectiveness in the First World War. In Allan R. Millett and Williamson Murray (Eds.), *Military Effectiveness: The First World War* (pp. 116-156). Cambridge University Press.

Rainey, James W. (1983). Ambivalent warfare: The tactical doctrine of the AEF in World War I. *Parameters: Journal of the U.S. Army War College 13*(1), 34-46. doi:10.55540/0031-1723.1327.

Rainey, James W. (1992). The questionable training of the AEF in World War I. *Parameters: Journal of the U.S. Army War College 22*(1), 89-103. doi:10.55540/0031-1723.1638

Seidule, Ty. (2014). Introduction. In Clifford J. Rogers, Ty Seidule, and Samuel J. Watson (Eds.), *The West Point history of the Civil War* (pp. 1-13). Simon and Schuster.

Stevenson, David. (2011). *With Our Backs to the Wall: Victory and Defeat in 1918*. Harvard University Press.

Trask, David F. (1993). *The AEF and coalition war making, 1917-1918*. University Press of Kansas.

United States War Department. (1899). *Annual report of the War Department for the fiscal year ended June 30, 1899*. Government Printing Office.

Vandiver, Frank E. (1977). *Black Jack: The life and times of John J. Pershing* (Vol. 2). Texas A&M University Press.

Weigley, Russell F. (1967). *History of the United States Army.* Macmillan.

10

Proximity and Distance on the Battlefield: The AEF's 2nd Infantry Division at Blanc Mont 1918

Keith D. Dickson
Joint Forces Staff College, National Defense University

Abstract

Proximity and distance were two important factors that shaped the role of the American Expeditionary Force (AEF) in 1917–1918. Heavily influenced by the orders and the ultimate goals of the AEF commander, General John J. Pershing, the AEF established both a proximity and distance to allied forces, which translated into a combat performance that, although effective, resulted in significant combat losses. The 2nd Infantry Division reflected the different ways that proximity and distance influenced actions on the battlefield, especially in the unique role the division played as part of the French Fourth Army in the attack on Blanc Mont in October 1918.

Keywords: Blanc Mont, Pershing, Lejeune

Proximity and Distance on the Battlefield: The AEF's 2nd Infantry Division at Blanc Mont 1918

Change is part of war's essential character. War is never static; the circumstances that make war so complex—fear, uncertainty, and chaos—create a requirement for continuous adaptations and new directions. The experience of the Great War, as terrible as it was, represented continuous technological innovation and the evolution of tactics, techniques, and command structures, especially on the Western Front. Although often perceived as a war of unrelenting, wasteful, and futile attacks, the battles of 1914–1916 gave way to increasingly innovative methods of employment of both troops and new technologies. In less than 3 years, changes emerged on the battlefield that heralded the onset of what would be recognized in less than a decade as modern operational-level warfare. Both the Allies and the Germans employed poison gas early in the war, initiated aerial bombardment and reconnaissance, and developed sophisticated complex logistics systems. By 1917, the Allies had begun to employ an entirely new combat system, the tank, integrated with artillery and infantry in combined arms attacks, supported by air power. The final defeat of the German army in 1918 came about through the effective employment of these forces. The Germans employed their own brand of innovation with the tactical use of storm troops, small unit infantry infiltration backed by precise artillery fire and gas. This innovation was the main reason for the initial and dramatic success of the spring offensive of 1918.

While the major combatants were demonstrating remarkable adaptations and employing forces on the battlefield with increased sophistication, the United States joined the war. The manner in which the Americans would respond to these changing circumstances, both operationally and tactically, was the responsibility of General John J. Pershing, commander of the American

Expeditionary Force (AEF), and his staff. The Americans, both intentionally and unintentionally, continuously maintained a position of proximity and distance to their allied partners. Pershing bluntly told Lieutenant General Sir Douglas Haig, the British commander, that he would not allow American troops to be subject to the will of the Allies. Although Pershing cooperated with General Henri-Philippe Pétain, the commander of the French armies, and Marshal Ferdinand Foch, the supreme Allied commander, he made it clear that he intended to form an independent American army. His absolute insistence on the Allies accepting this condition of American participation, along with his refusal to accept the hard-won battle experience of the French and British as a legitimate source for preparing his own divisions, created an often frustrating and awkward sense of distance even while maintaining an association of a unified allied effort. (Smythe, 1986, p. 171, 176-177) These combined factors of proximity and distance in military operations with the arrival of the Americans into the war can be understood through the following aspects:

Command relationships. Pershing's agenda, based on his orders from the secretary of war as well as his own views of how to win the war, created an enduring distance between the AEF commander and the Allied command. American units would cooperate with the Allies and establish liaison and coordination, acting as close to Foch's intent as possible, but at the same time maintain a separate zone of action for an American army under American command and fight the enemy in its own uniquely American manner. The 2nd Division also had a unique command situation that lent itself to a simultaneous condition of proximity and distance. It was an Army unit under the command of a Marine. General John A. Lejeune followed Pershing's lead as a subordinate to French commanders, refusing to allow the division to be broken up and placed under direct French command, and he adopted the American style of open warfare instead of conforming to the French tactical approach.

Battlefield coordination. Allied divisions shared boundaries, identified on maps and transmitted in orders, which were intended to enhance the coordination and control of forces in maintaining rates of advance during an attack. The proximity of large units next to each other required coordination to prevent units from straying into another division's area and firing on each other by mistake. The Americans, while in proximity to French units, often advanced without regard to boundaries or the need to maintain contact with units on their flanks. Consequently, American units often found themselves isolated and exposed to flanking fire from enemy units. The 2nd Division prided itself on its esprit de corps and its reputation as a first-rate assault unit. Leaders and troops had little regard for any French unit that did not match its aggressive style of combat and complained continuously of French units failing to keep up the rate of advance.

Proximity and distance to the battlefield. American divisions were constantly being rotated in and out of the battle area or being transferred from one portion of the battle area to another. The Americans often found themselves marching or transported long distances with little time to rest or refit and being moved directly into the attack. This continuous movement between proximity and distance that the AEF underwent from late 1917 and through much of 1918 was mirrored in the battle record of the 2nd Infantry Division. This division gained more ground and had more days in combat than any other American division. It was one of the best-equipped divisions and participated in the first American-led offensives at Soissons and St. Mihiel, where it operated in conjunction with French divisions.

Tactical employment. The American units accepted French training reluctantly, enduring the exposure to trench warfare in quiet sectors under French supervision. The Americans also chafed under French instruction on combined arms tactics, preferring their own brand of tactical employment that Pershing dictated—infantry attacks intended to move from the static and confining limitations of the trenches to what he called open warfare. While maintaining a physical proximity to the French on the battlefield, the Americans simultaneously were seeking to distance themselves from what they believed to be a tactical approach that brought no decisive result. The 2nd Division experienced an internal dynamic of proximity and distance that was unique in the AEF. The division had two brigades—one Marine and one Army. The Marines had been foisted upon the Army at the insistence of the commandant of the Marine Corps, who wanted Marines playing a prominent combat role. In the division's combat employment, the Marines were often heralded in the press at the expense of their Army comrades, who resented that their battlefield courage and accomplishments were not equally recognized. Service rivalries never faded, and Marine and Army officers often clashed. This condition of proximity and distance resulted in often unorthodox tactical employment of the brigades—leaving each brigade to fight separately rather than in close coordination.

To gain an appreciation for the significance of this simultaneous condition of proximity and distance and how it influenced the battlefield performance of the 2nd Infantry Division, it is necessary to examine each of these four aspects in terms of how Pershing's influence shaped the way in which his subordinates followed his requirements and expectations.

Command Relationships

After being designated as the commander of the AEF, Pershing received clear orders from Secretary of War Newton Baker:

> You are directed to cooperate with the forces of the other countries employed against the enemy, but in doing so, the underlying idea must be kept in view that the forces of the United States are a separate and

distinct component of the combined forces, the identity of which must be preserved. This fundamental rule is subject to such minor exceptions in particular circumstances as your judgment may approve. (Goldhurst, 1977 p. 260)

These orders set the stage for the resultant simultaneous proximity and distance that Pershing established with the Allied command and the Allied political leadership. The French and British sought to integrate American troops into existing units allied as wholesale replacements (the ultimate proximity), essentially subsuming the Americans into an anonymous levy. Pershing possessed a stubborn conviction that the American contribution to the war would be the decisive factor for total victory. Pershing assessed that the French and British armies were near culmination, unable to do much more than hold out against increasingly powerful German forces. Despite an obvious lack of combat skills, equipment, and experienced combat leadership, Pershing believed the American fighting spirit would turn the tide of battle in favor of the Allies. For this fighting spirit to be decisive, however, Pershing insisted that the American army had to be completely independent and an equal partner with the French and British. Even though Marshal Joffre, Pershing's liaison to the French General Staff, insisted that American officers and noncommissioned officers should be trained in French schools, largely in preparation for integration into French units, Pershing resisted any efforts to use newly arriving American troops as replacements on the front lines.

Instead, he built a separate American force of divisions capable of winning the war, not by attrition warfare and stalemate of the trenches, conditions which he believed had sapped the will of the soldiers to fight. Pershing intended that the American army would be employed as an assault and maneuver force. Bayonet and rifle fire were the decisive means—energy and aggressive spirit were needed to overcome the inertia of trench warfare. While he tolerated the French and British tutelage in exposing American units to the trenches in quiet sectors of the line, Pershing emphasized large-scale unit maneuver training and bayonet fighting for the five divisions under his command. Pershing insisted that Americans take a completely different approach to battlefield tactics than what he believed the French and British had demonstrated up to 1917. To Pershing, individual initiative and rapid maneuver were the essential aspects of his concept of mobile warfare. An aggressive infantry assault employing effective rifle fire was more important in an attack than was supporting arms, such as tanks and artillery.

Pershing's insistence on training and preparation of a separate American force before committing them to combat caused a great deal of frustration and resentment for the French and British. Pershing maintained his distance from the Allies, who continued to fight alone while the Americans siphoned off resources and took almost 6 months to build combat readiness. The proximity of tens of thousands of American troops near the front lines waiting to be employed, but untouchable, certainly contributed to the Allies' concerns that the war very well could be lost before the Americans ever got involved.

The postwar history of the 2nd Division captured the distance in attitude that separated the Americans from their French counterparts. The history asserts that Pershing and his staff

> were in a position to study and appreciate the tactical methods they saw in vogue . . . They refused to admit that the nature of warfare had suddenly undergone a complete change . . . The foreigner, not knowing the military background and competence of our leading minds, looked upon us as arrogant and ignorant novices, too self-centered to weigh advice. (Spaulding and Wright, 1989 p. 9)

The Americans looked on the "doctrine of the trench defensive" as detrimental to "an aggressive offensive and training in open warfare" (Spaulding and Wright, 1989, p. 9).

The Americans were quite skeptical of the British and French doctrine formed by their years engaged in trench warfare. This static warfare, the Americans believed, had blunted their offensive spirit, making the advance of the trench line the objective of battle. The Americans, who drew deeply from the lessons of the Civil War, believed that success was reliant upon aggressive offensive maneuver and open warfare (Mead, 2000 p. 182-183).

The 2nd Infantry Division was different from other divisions in the AEF. Although it had the same number of men (28,000), it had a brigade of about 8,500 Marines. The 2nd Division was not organized and deployed from the United States as were other AEF divisions; it was assembled from units already in France. The division headquarters was established in France in October 1917, and, after it became operational in late January 1918, the division began a 3-month training plan that involved 1 month of small unit training, followed by battalions spending a month occupying a quiet sector of trenches under French supervision, followed by full division-level training. The 2nd Division history notes that the instruction provided by the French created problems, described as "difference in language, temperament and methods" and prevented the Americans from gaining very much from their training (Mead, 2000 p. 7-8). "Many defects in this system were perceived," the division history admitted, "but nothing better could be worked out practically" (Mead, 2000 p. 7-8). Despite their exposure to the French tactical instructors and their knowledge of the realities of the battlefields of 1917, the troops and leaders of the 2nd Division reflected Pershing's intent to maintain a formal distance from their allies, reflecting the widely shared American view that the French were only interested in maintaining a trench warfare capability.

Battlefield Coordination

About 100 German divisions were being transferred from the Eastern Front, in the aftermath of the collapse of Russia, to the Western Front for the final offensive that would win the war. German storm troops shattered the Allied lines and led to the supreme crisis in the war. The German spring offensive in March and April was intended to defeat the Allied forces in detail with a main attack against the British to break the defensive line in northern France and drive the remnants back toward the Channel coast; the attack appeared decisive as German forces advanced 60 km through the British lines. In May, a series of supporting attacks against the French were launched to prevent any attempts to reinforce the British. The French employed 16 divisions to block what was perceived to be a thrust to capture Paris. In this crisis, the British and French turned to the Americans—ready or not, American forces had to be committed.

Pershing, responding to the appeal, had the authority from Baker in his orders to use his judgment to support the Allies in special circumstances. He began to parcel out some of his divisions. The 42nd Infantry Division initially joined the British, as did the 27th and 30th Infantry Divisions shortly thereafter (these latter two divisions served under British command for the remainder of the war). In a series of major offensives, sequenced with more limited thrusts intended to split the British and French armies, the Germans spent their last resources. Unable to secure a rapid victory, they would now hold out as long as possible until negotiations brought an end to the fighting on favorable terms to Germany.

By late May, the 2nd Division was alerted and within 12 hours had begun to march forward in the general direction of the Château-Thierry-Paris road to blunt the massive German offensive. Here the division, in nearly continuous combat from May 31 to July 25, gained a reputation as an effective assault division. The 2nd Infantry Division was unique in that it had one brigade of two Army infantry regiments (the 9th and 23rd Infantry) and a brigade of two Marine infantry regiments (the 5th and 6th Regiments). It was an awkward pairing that created an odd blend of proximity and distance based on service traditions and cultures. The Marines were volunteers, generally better trained and physically prepared for battle than were the Army soldiers. They were proud to be a force apart and tended to act independently, exhibiting a certain disdain for the French as well as for their Army counterparts.

For their achievements in Belleau Wood, the Marines came to be known by the Germans as *Teufelshunde* (Devil Dogs). This first contact with the enemy was especially a battle of open warfare—bayonet attacks against dug-in German defenders. In July, the division was again sent into combat with little prior warning and trucked to an unknown location from which it spent a full day marching to the front lines and initiating an attack at Soissons. All of these attacks followed the standard practice of successive infantry assaults, with varying success at coordinating supporting arms, especially artillery and machine guns. Over the next few weeks, the results of these tactics were telling: 1,900 total casualties at Château-Thierry; 3,900 casualties at Soissons; at Vaux, the division suffered 7,500 casualties. The Marines alone lost 4,600 men during this time. In September, the division participated in the St. Mihiel offensive, the first time the American army fought as an independent command. The division distinguished itself with the 9th and 23rd Infantry, capturing Thiaucourt along with 3,000 prisoners and 92 artillery pieces. At St. Mihiel, the division lost 1,500 men (Spaulding and Wright, 1989 p. 132, 160; McClellan, 2014 p. 115).

Both the Allies and the Germans looked at the American effort in the same way. They were filled with admiration for the courage displayed and the aggressive spirit that carried the troops forward but saw it as a needless sacrifice of good men. The Americans had difficulty making proper coordination with machine gun units and artillery; any attempts at coordination with flanking units was abandoned early. They often displayed contempt for the French units with them, and, once an attack was launched, everything became improvisation. Any control and order above the company level was usually lost, and troops found themselves fighting for days without food or water or any type of resupply. Replacements were assimilated rapidly, but small unit leaders, both officers and noncommissioned officers, were lacking in tactical skill, and many were largely untrained. The division itself changed in its makeup of tactical components as the ratio of veterans to new replacements shifted continuously. Marine and Army officers often did not cooperate, leaving each brigade to fight its own separate battle instead of developing a coordinated division attack.

Army Brigadier General James G. Harbord, Pershing's chief of staff, took command of the 4th Marine Brigade prior to the division entering combat. At Belleau Wood, Harbord followed Pershing's tactics to the letter, resulting in terrible casualties but gaining a measure of immortality for the Marines, largely due to his close association with the press. Harbord was promoted to Major General and took command of the 2nd Division. He continued his blind allegiance to mobile warfare, committing the division in his unimaginative orders to unrealistic and unsupported attacks that led to serious losses and left the division largely combat ineffective. Commanding the division for only a month, Harbord was reassigned to lead the logistics support effort as commander of the AEF's Service of Supply. In late July, Major General Lejeune became division commander—a Marine leading an Army division was unusual in the extreme. Lejeune immediately noted that "the personnel [structure] of the Division was in a constant state of change" (Lengel, 2015 p. 204-205, 288; Lejeune, 1930 p. 334).

The 2nd Division had fought magnificently and had accomplished extraordinary feats through raw courage and sheer determination. However, this was not a substitute for careful planning and coordination, and the results and unnecessary losses were staggering. Unfortunately, the American success in halting the German offensive and then going on the offensive seemed to confirm that the American approach of open warfare, despite the heavy casualties, was valid. The pattern was now set: American combat troops were committed to infantry assaults—despite enemy artillery and machine guns—to prove to the Allies that this style of battlefield employment would bring victory. Even within the division itself, however, proximity and distance existed simultaneously. The Marine brigade and the Army brigade functioned separately, often failing to support each other. Each brigade also had difficulty in employing attached machine gun units or coordinating supporting artillery fire once the attack had begun. Replacements for casualties weren't as well trained, and leaders had little time to make adjustments based on battlefield experience.

By August 1918, as the initiative passed to the Allies, Pershing indicated he was ready to employ the American army. Marshal Foch presented to Pershing a plan for a limited offensive against the St. Mihiel salient, while the bulk of the American forces supported the French Second Army's attack toward Mézières. Pershing insisted that only an independent American army would be employed. After a long and acrimonious debate, it was agreed that Pershing would have his wish. A separate and independent American army under his command would conduct a two-stage offensive. In the first phase, the Americans would participate in the reduction of the St. Mihiel salient; with that accomplished, Pershing would reorient the US First Army northwest to attack through the Argonne Forest toward Metz (Lengel, 2015 p. 25, 63, 276).

Proximity and Distance to the Battlefield

The 2nd Division was often moved in and out of the forward lines on short notice. The officers and men had little idea where they were, where they were going, or how close to the front they actually were. There was a constant and continuous shifting of proximity and distance to the battlefield itself that influenced the employment of the elements of the division and how orders were transmitted. Rail transport was especially effective in bringing troops from a distant location to the proximity of the battlefield in a relatively short time. After the battle of St. Mihiel, the division was relocated to Toul in September. When the 2nd Division was assigned to the French Fourth Corps, a rail movement was ordered—but no destination was specified. The movement nonetheless lasted 3 days, from September 25 to 28. Unfortunately, the division was left to deal independently with the French railway system, with both the trains and crews under French control. The movement of an American division required 58 trains—the equivalent rail transport required for two French divisions. Packed into the train cars, the troops were never sure where they were, when they would arrive, or where they were going. The continuous shifting from an assembly area located at a distance from the battlefront back to the battlefront in a matter of hours also had a significant effect on the combat capability of the brigades in the division. New soldiers and Marines arrived by the hundreds daily to replace combat losses in the brigades, but there was rarely time between coming out of one battle and entering the next that replacements and leaders at the platoon and company level could receive any significant training or preparation. American leaders also tended to forget experiences in battle that demonstrated the importance of supporting arms; instead, they continued to revert to wave assaults despite suffering the effects of such tactics in previous battles.

Tactical Employment: Blanc Mont

Marshal Foch, after a series of smaller offensives, prepared for a concerted Allied effort to break the German armies. The US First Army, in conjunction with the French Fourth Army on its left, were to undertake a combined offensive. The American army, bounded by the Meuse and the Argonne Forest, would make a general attack along a 12-mile front with 200,000 men on September 26 and October 4 to clear German strongpoints and outflank the defenses. The French Fourth Army advanced northwest to break the German main line of resistance, seize the strongpoint at Blanc Mont, and advance toward the Aisne River. The Fourth Army sought an initial breakthrough and exploitation, and, on September 16, General Pétain requested three American divisions to strengthen the attack. Pershing provided two divisions, the 2nd and the 36th Infantry. The 2nd Division, replenished with replacements, was at full strength and was dispatched to the French army sector to be held as part of the general reserve (Trask, 1993 pp. 138-139). As a separate division under French command, it played a significant role as part of the French Fourth Army's attack in October of 1918. Yet the same issues of proximity and distance that characterized the AEF continued to be displayed in the 2nd Division.

Lejeune endorsed what he described as General Pershing's concept of the "vital importance" of open warfare (Lejeune, 1930 p. 291). Lejeune's own words describing the Belleau Wood battle indicate that he shared Pershing's view that American courage and aggressive spirit made the difference, despite the human cost:

> The losses were heartbreaking, but the determination and the valor displayed by these young men [of the 5th and 6th Marine Regiments]

electrified the world . . . and it became known to all men that the Marines had made a successful attack on the formidable foe . . . Over five thousand Marines were killed or wounded within [Belleau Wood's] borders. It is holy ground. (Lejeune, 1930 p. 294-295)

General Lejeune met with General Henri Gouraud, commander of the Fourth Army. As elements of the division were arriving in their assembly areas, Lejeune learned that the French were intending to divide up the division and assign individual brigades to augment French divisions. Lejeune appealed to Gouraud, who indicated that the Fourth Army's forward movement had been stalled by a German defensive position in the center of the Army's avenue of advance that marked the enemy's main line of resistance. Lejeune described it as "a powerfully fortified range of hills," which he indicated were "too strong to be carried by direct assault." This was Blanc Mont. The capture of Blanc Mont would force a general retreat and cause the Germans to abandon their attempts to capture Reims. Gouraud indicated that the French units were exhausted and could not continue the offensive without significant reinforcement from the American brigades (Lejeune, 1930 p. 341; Mead, 2000 p. 314).

Following Pershing's approach of simultaneous proximity and distance, Lejeune sought to preserve his division but also support Gouraud. Lejeune proposed that the 2nd Division be brought to the front line and attack toward Blanc Mont on a narrow front to create a breakthrough, supported on its left and right flank by the French divisions. "I am convinced," he told Gourand, the 2nd Division "will be able to take Blanc Mont Ridge, advance beyond it, and hold its position there" (Lejeune, 1930, p. 342). Lejeune's option was approved, and the 2nd Division was placed under the operational command of the French Fourth Army, with orders directing the division to move to the front lines soon following.

On October 1, the 2nd Division relieved the remnants of the French 61st Division at the front, whose companies had been reduced to platoon strength after an advance of 5 km to the Essen trench line, about 3 km short of Blanc Mont, rising 200 ft above the valley. This dominating terrain was the defensive key to the Champagne region. The capture of Blanc Mont would break the German main defensive line and allow the entire French XXI Corps to advance 30 km to the Aisne River, the next defensible position for the Germans. The 2nd Division, as part of the French XXI Corps under the command of Major General Stanislaus Naulin, held the far left of the Corps boundary with the French XI Corps. The Americans occupied a 2-mile front. The old German trench system was 500 yd deep and consisted of four trench lines. Essen Hook, located about 500 yd west of the division flank, was strongly fortified, with concrete machine gun positions that dominated the flat open ground to the front of the Americans. The terrain beyond was broken by a series of hills and ridges, interspersed with ravines, depressions, and small patches of woods. It was described as "rolling, with fairly steep slopes, and covered with patches of scrub pine; these along the front lines, had been devastated by shell fire." The ground was composed of "chalky limestone, showing clear and white wherever the surface is torn up" (Van Every, 1928, p. 169).

It was ideal terrain for the defense, especially for an enemy force that had perfected the use of the machine gun as a defensive weapon, covering all open terrain with interlocking and mutually supporting fields of fire. The experienced German soldiers of the 200th Infantry Division also could skillfully employ the machine gun as an offensive weapon, moving and reemploying these weapons to support counterattacks or moving them to engage enemy flanks. Supported by artillery and mortar fire, the infantrymen could move along the hills and ridges through a maze of trenches, covered ways, bunkers, and observation posts, allowing them to shift forces to meet attacks from nearly any direction.

The 2nd Division attack plan was a tactical curiosity. To maintain the momentum of the assault, the two brigades purposely avoided a German strongpoint in the center of the division sector, the Bois de la Vipère. The 4th Brigade (Marines) would advance on Blanc Mont from the left, while the 3rd Brigade (Army) would attack on the right. The overall intent was to have both brigades converge simultaneously on the objective. The drawback to this approach was that the two brigades would be separated by nearly a kilometer, preventing any mutual support. In addition, by ignoring the strongpoint in the initial attack (it was to be taken after Blanc Mont had been captured), the brigades were subject to flanking fire during the attack, as well as being potentially cut off with a substantial force now located behind them. This intentional distancing of forces in the face of the enemy, especially in the broken terrain of the 2nd Division's battle area, only further contributed to the isolation of American units and diluted the power of the attack. The American concept was summarized in the words of one officer: "We'll make a hole and broaden it later" (Van Every, 1928 p. 287).

With the two brigades essentially punching a hole by converging on a specific point, coordination along the division boundaries would be very difficult, if not impossible, creating a dangerous distance between the French divisions on the flanks. This situation, in turn, prevented General Naulin from directing any type of coordinated corps-level advance. Instead, each division would have to fight its own individual battle within its boundary. For the 2nd Division, which had advanced farthest, this meant that its flanks would most likely be continuously exposed to fire from German defenses across the divisional boundaries.

Lejeune changed the attack schedule to October 3 to bring up supporting artillery, allow units to observe the terrain they were going to attack, and allow for forward units to clear the trench lines of Germans so that there would be a clear line of advance. The XXI Corps order arrived at 2nd Division headquarters at 11:00 p.m. for a scheduled 5:00 a.m. attack on October 3. The order had to be translated and studied, and then a division order had to be prepared, reviewed, and issued to the brigades, who received it at about 4:00 a.m. The brigade headquarters had to prepare an order that, in turn, was passed to the subordinate battalions. The companies received the battalion order, which then had to be refined for the platoons. Until orders were passed, units were in a state of flux—in many cases, a simple set of verbal orders was given with almost no time for preparation. In one battalion, the commander received the brigade order at the time of the attack and did not read the orders until his unit had reached the objective (Lejeune, 1930, p. 351; Spaulding and Wright, 1989, p. 170).

The division would follow an axis of attack 2.5 miles wide, with each brigade attacking in columns of regiments, and each of the regiments' three battalions advancing in lines. The leading battalions of each brigade would be accompanied by a company of 12 French tanks. The battalions would have machine gun companies in direct support accompanying the attack. A mix of French and American artillery would lay down a short but intense preparatory barrage just before H hour (5:00 a.m.—the time specified by the Corps order). The 9th Infantry Regiment would lead the 3rd Brigade attack, which was to advance northwest to outflank Blanc Mont and provide support to the 4th Marine Brigade. The 6th Marine Regiment would lead the 4th Brigade attack to seize Blanc Mont. The dividing line between the brigades was the Somme-Py-St. Etienne road. The ultimate division objective was the Médéah Farm-Blanc Mont road, 2 miles from the jump-off point. The French 167th Division on the right flank and the French 21st Division on the left flank would advance alongside the Americans.

Advancing in a dense formation over open ground, the Americans found themselves subjected to a maelstrom of artillery and machine gun fire. The tanks assisted in destroying some machine gun positions but lagged behind the infantry. Despite heavy casualties, the American brigades had reached a general objective line between 8:40 and 9:00 a.m. The 3rd Brigade had reached the Médéah Farm, and the 4th Brigade occupied the face of Blanc Mont. Despite their losses, the Americans had driven a 4.5 km salient into the German lines, but it was only 500 yd wide at the apex, and German forces were still operating freely in the center of the salient and on the hills between the separated American brigades. German machine gunners infiltrated the gaps, forcing the Americans to halt any forward advance and curl back to defend their open flanks from enemy fire. The French divisions had failed to keep pace, and the Americans were exposed to heavy flanking fire, especially at the boundary between the French and Americans on the left flank at Essen Hook and from the western part of Blanc Mont Ridge, all within the 167th Division's boundary. The 3rd Brigade at Médéah Farm was halted and suffered heavy casualties from German positions across the boundary that were the responsibility of the 21st Division. Although ordered to make a follow-on attack, the 23rd infantry and 5th Marines could make no significant progress. Lashed by artillery and machine gun fire from their open flanks, they could only make a weak defensive line south of the Médéah Farm-St. Etienne road. The French were still far below Médéah Farm on the right and south of Blanc Mont on the right.

The two brigades, separated by a German strongpoint, continued the attack throughout the day, suffering heavy casualties. Units became isolated in the low scrub and shallow valleys, and the battalions were forced to confront enemy forces to their east and west instead of advancing north. The Marines of the 4th Brigade, instead of continuing the attack, were forced to cross the divisional boundary and clear the Essen Hook, a trenchline occupying terrain that overlooked the 2nd Division sector and which allowed the German defenders to use machine guns to pour a continuous high volume of accurate fire against the brigade (Van Every, 1928, pp. 283-286, 352-354, 357; Otto, 1930, pp. 115-119).

On October 4, the outflanked and increasingly isolated brigades defended their tenuous gains against heavy enemy artillery bombardment and intense infantry counterattacks while waiting for the French divisions to advance. The commander of the 9th Infantry refused to advance until the French came up to protect their flank. By October 5, the French divisions had advanced sufficiently to eliminate German flanking fire. After a 30-minute artillery preparation, the 6th Marine Regiment cleared the German forces still defending the western slope of Blanc Mont and linked with the French division on its left. The 3rd Brigade was initially unable to advance until relieved of the threat to its open flank. It later attacked north before being stopped by strong resistance 2 km south of St. Etienne, close to the division objective. As the American brigades linked and advanced toward St. Etienne, the Germans made an adroit retreat to a pre-planned second line of resistance, having met the goal of inflicting the greatest damage on the attacking force as possible while preserving combat strength. The German units had been severely damaged over the 3 days of combat but maintained enough cohesion to break contact and withdraw while inflicting heavy casualties on the exposed American troops. On October 6, both brigades attacked to capture St. Etienne but were unable to advance against the German strongpoints. Between October 6–7, the 2nd Division was partially relieved by elements of the US 36th Infantry Division, which had never been in combat before. By October 10, all the elements of the 2nd Division had been replaced on the front lines and were moved to a rear area (Historical Branch, 1922, pp. 4-8, 11-14; Quinn, 1933 pp. 10, 12-18; Schubert, 1941 pp. 112-113).

From the perspective of the French high command, the attack on Blanc Mont was a resounding success. The Germans held St. Etienne for only a short time before being forced to retreat to the Aisne River. It was one of the great offensives of the war, guaranteeing that Germany would lose the war. General Gouraud delivered a letter to General Pétain, declaring that the 2nd Division "played a glorious part in the operations of the Fourth Army" (Lejeune, 1930, pp. 365-366). The division was cited in orders of the French Army and over 2,000 Croix de Guerre were awarded.

From October 2–10 at Blanc Mont, the 2nd Division had lost nearly 5,000 dead and wounded—a shocking casualty rate for so short a time in combat but nevertheless typical of the losses the division had suffered in previous battles. Lejeune's tactical plan of an attack of brigade columns separated by nearly a mile, while bypassing a major enemy strongpoint, guaranteed heavy losses but would gain ground, as he promised Gouraud. Unlike most other divisions, the 2nd Division coordinated with the French units on its flanks, had French tanks advancing with the infantry, had machine gun companies in direct support of the battalions, and had French and American artillery provide an initial barrage on enemy positions. On the surface, it appeared that Lejeune and his staff had adopted the French and British method of combined arms attack. This was not the case, as the course of the battle illustrates. While French units were in proximity to the division, the Americans immediately distanced themselves from the French, advancing at a rate intended to create a breakthrough, while the French divisions on the division's left and right made a more deliberate advance. The result was predictable. As had happened in previous battles, the exposed flanks allowed the Germans to halt the advance and force the Americans to fight in multiple directions. Separated and isolated, the American brigades could not advance until the French divisions caught up. The French tanks are not mentioned in the battle reports of the brigade commanders and disappear from any orders. They may have had some use in suppressing German machine gun positions in the initial attack but apparently played no role after October 3. The inability or unwillingness to coordinate with the French tank companies limited the effectiveness of the American attack, forcing the infantry to face entrenched infantry and machine guns alone. Scheduled attacks were always preceded by artillery fire, but once the attack started, sustained fire on enemy positions in support of the infantry was nearly impossible due to a lack of communication. The two brigades did not fully link up until nearly 2 days into the battle, and the men fought on with little rest and no food or water.

The Americans never appreciated what the French had attempted to teach them: the battlefield of 1918 demanded a deliberate advance, using coordinated artillery fire and tank support. Tactical success required a careful and highly orchestrated employment of supporting arms in support of infantry. Carefully planned attacks could be successful against strong defenses if aircraft, tanks, and artillery all worked in close coordination. Unfortunately, it was a pace and an approach the Americans could not accept. The result was almost always the same: French units advanced far more slowly, and the rapid American assault often left the units with open flanks subject to enemy fire from three directions. Command and control was often lost and there was little coordination with the machine gun companies attached to the regiments (Griffith, 1994, pp. 193-200).

Proximity and Distance: Two Defining Factors

General Pershing set the tone for how the AEF would be employed in combat. His expectations were that a separate American army would have enough proximity to the French and British to support the Allied broad strategy and Marshal Foch's operational guidance for the grand

Allied offensive, but he would establish an immutable distance between the Allies and Americans in terms of combat methods and employment of forces. Despite terrible losses, Pershing could not abandon his position for fear of losing control of American forces. Pershing found that he had to compromise on his absolute control of American forces, essentially loaning several American divisions to the British and French to support Foch's offensive. The 2nd Division was prominently involved, finding itself more or less in constant motion from one portion of the battlefield to another and from one command to another. The division was in a state of constant flux; it was being not only continuously renewed with thousands of minimally trained replacements but also shuttled into and out of the battlefield—a process of proximity and distance that disoriented leaders and troops and continued for the 2nd Division for the duration of the war.

Pershing's insistence on open warfare, aggressive infantry assault relying on rifle fire and the bayonet, were in complete contradiction to the emerging tenets of combined arms operations that marked Allied offensives in 1918. American divisions, especially the 2nd Division—one of the most effective combat divisions in the AEF—followed Pershing's dictum. While performing with splendid courage, the two infantry brigades suffered very high casualties in relation to results achieved. General Lejeune, like Pershing, resisted breaking up the division when attached to Gouraud's Fourth Army. Instead, he committed the division to a frontal breakthrough assault on one of the most formidable defensive positions in the Fourth Army's area of responsibility. Americans would distance themselves from the French approach with a purely American style of attack. The division's achievement at Blanc Mont was impressive, and, arguably, only American Marines and soldiers could have captured that formidable ridge line. By maintaining a distance from the French tactical approach, Lejeune laid out a plan of attack that, while on the surface appeared to conform to the French model, almost immediately abandoned that model and became a pure infantry assault. The Americans could not and would not adapt to the battlefield realities. The essential proximity of French flank units was the only real means to success, but the Americans ignored the need for a steady, coordinated advance using supporting arms. The spirit of the offensive was everything; nothing was to stop the advance and the breakthrough. As a result of this separation of physical distance on the battlefield, the Americans did indeed advance, did indeed break through, and won immortal glory, but at a punishing cost.

References

Goldhurst, Richard. (1977) *Pipe Clay and Drill John J. Pershing: The Classic American Soldier.* New York: Thomas Y. Crowell Company.

Griffith, Paddy. (1994) *Battle Tactics of the Western Front: The British Army's Art of Attack 1916-1918.* London: Yale University Press.

Grotelueschen, Mark E. "The Doughboys Make Good: American Victories at St. Mihiel and Blanc Mont Ridge." *Army History* (spring 2013): 7-18.

Lejeune, John A. (1930) *The Reminiscences of a Marine.* Philadelphia: Dorrance and Company.

Lengel, Edward G. (2015) *Thunder and Flames: Americans in the Crucible of Combat, 1917-1918.* Lawrence, KS: The University Press of Kansas.

McClellan, Edward N. (2014) *The United States Marine Corps in the World War*, 3d ed. Washington, D.C.: U.S. Government Printing Office.

Mead, Gary. (2000) *The Doughboys: America and the First World War.* New York: The Overlook Press.

Nelson, James Carl. "The First Day at Blanc Mont." *Leatherneck* (November 2016): 12-15.

Nenninger, Timmothy K. "Tactical Dysfunction in the AEF, 1917-1918." *Military Affairs*, 51, no. 4 (October 1987): 177-181.

Otto, Ernst. (1930) *The Battle at Blanc Mont*. Annapolis: U.S. Naval Institute.

Quinn, Lawrence A. "Critical Analysis of the Infantry Scheme of Maneuver of the 2d Division at Blanc Mont in October 1918." (1933) Ft. Leavenworth, KS: Command and General Staff School.

Schubert, R. H. "A Critical Analysis of Flank Protection, Second Division (A.E.F. France) 3 October to 9 October." *Marine Corps Gazette* (November 1941): 112-113.

Simmons, Edwin H. "With the Marines at Blanc Mont." *Marine Corps Gazette* (November 1993): 34-43.

Smythe, Donald. (1986) *Pershing: General of the Armies*. Bloomington: Indiana University Press.

Spaulding, Oliver L., and John W. Wright. (1989) *The Second Division American Expeditionary Force in France 1917-1919*. Nashville, TN: The Battery Press.

Trask, David F. (1993) *The AEF and Coalition Warmaking, 1917-1918*. Lawrence, KS: University Press of Kansas.

Van Every, Dale. (1928) *The AEF in Battle*. New York: D. Appleton and Company.

War Plans Division General Staff, Historical Branch. (1922) Monograph No. 9 "Blanc Mont (Meuse-Argonne-Champagne)," June 1921. Washington: U.S. Government Printing Office.

11

AEF Press Censorship During World War I

Charles Sorrie

Abstract

This chapter argues that American press censorship practices in Europe were both successful and necessary during World War I. The primary functions of military press censorship were to safeguard secrets and to shore up both military and domestic morale. The American military's bureaucratic framework for censorship, after establishing its foundations in 1917, matured throughout 1918 in response to the United States' increased military presence in France. By the time the Doughboys were engaged in heavy fighting in the fall of 1918, the efficacy of America's information management system in Europe matched that of its Allied counterparts. Though criticized immediately after the war, America's successful World War I military censorship program later provided a model upon which to be built during World War II and Korea.

Keywords: censorship, World War I, military press

AEF Press Censorship During World War I

The American Expeditionary Force (AEF) during the First World War developed America's first centrally administered bureaucracy for military press censorship. The Press Censorship Division, Military Intelligence Section (G2D), like its Allied counterparts, was primarily designed to protect military secrets. Also, like the British and French censorship systems, it engaged in counterintelligence activities and coordinated with domestic and foreign propaganda agencies. In June 1917, the relationship between General Head Quarters (GHQ), the War Department, and America's first journalists and censors in Europe was strained. During the following 14 months, however, tensions between these groups gradually abated. By fall 1918 these groups together administered an information management system built on mutual trust and interdependence. By the outbreak of the Meuse-Argonne offensive in late September, the AEF had developed a bureaucratic mechanism able to collect and distribute news covering large military operations spanning vast distances. When it was needed the most, G2D was ready to protect both American and Allied soldiers from security leaks and to help shore up the home front.

This chapter was primarily researched using the rich archival collection on G2D's wartime foundation, correspondence and operations located in the National Archives and Records Administration's AEF collection (Record Group 120). Other archival materials were collected from two Parisian archives. The *Service Historique de la Défense* contains the French War Ministry's records (5N). The French Press Bureau was directed by the War Ministry and much important correspondence between it and the AEF regarding press censorship is found in this collection. The *Bibliothèque de Documentation Internationale Contemporaine* contains an extensive and highly accessible collection on French wartime censorship (F rés 0270). It includes a particularly useful set of registries containing orders from the French Press Bureau to French censors and editors, to the AEF, and to all foreign owned newspapers operating in France. This international approach to the research of American censorship during World War I is unique within the historiography of the subject.

Little has been published on American military censorship during World War I. The subject has been discussed in subsections of larger studies on American security issues or wartime correspondents. Most literature on American journalism in France has focused on correspondents' personal experiences. Emmet Crozier's *American Reporters on the Western Front, 1914-1918*, published in 1959, was one of the first studies to use the National Archives' G2D collection. His archival findings are reinforced with memoires and interviews conducted with ex correspondents. It remains an essential secondary resource and has significantly influenced subsequent historiography of the subject. More recently, Professor Steven Casey, a leading 20th century American media historian, has extensively researched the subject. His generously provided unpublished manuscript was both interesting and highly useful in helping to navigate NARA's extensive AEF records.

Although G2D's practices were publicly debated during the 1920s, the historiography of American censorship practices during World War I has focused on the domestically repressive practices of the Woodrow Wilson Administration. There is a wide consensus that American military censorship was more severe than during subsequent twentieth century conflicts and this chapter's objective is not to moralize about the limitations of speech freedoms in wartime but to demonstrate how the AEF and its accredited journalists evolved through shared wartime experiences and worked together to contribute to the Allied war effort. Hopefully this study will lay a groundwork for media and military historians to further explore the subject.

Building G2D from Scratch

The AEF's Chief of Intelligence (G2) was Colonel Denis Nolan. He was appointed shortly after America joined the war and held the office until after the armistice. The Press and Censorship Division was one of four administered by the AEF's Military Intelligence section.[1] Nolan had little involvement with the daily activities of press censorship and primarily acted as an administrator and liaison to both GHQ and Washington. His officious, academic personality, and his experience in the Spanish-American War made him well-suited to the position. He soon appointed Major Marlborough Churchill to head G2D. Churchill was responsible for supervising G2D's bureaucratic organization and reported directly to Nolan.

Whereas G2D's top two administrative positions were manned by experienced army officers, a civilian, Frederick Palmer, was appointed Chief Press Officer. During the AEF's first seven months in France, Palmer organized the new press censorship office in Paris, oversaw the daily functions of press censorship, enforced the rules and regulations handed down by GHQ and supervised American correspondents. Palmer seemed the perfect man for the job. He was a friend of General John Pershing's (Commander of the AEF) and was a highly respected news reporter with 20 years of experience. He had been the only neutral correspondent accredited with the British early in the war and was accredited with the *Associated Press* (AP), the *United Press* (UP) and William Randolph Hearst's *International News Service* (INS). He would find the position both thankless and exhausting.

On May 28, 1917, Pershing, along with 40 regular army officers, 17 reserve officers, 67 enlisted men and a group of civilians, interpreters and clerks, boarded *The Baltic* in New York. Amongst the group were Palmer,

---

1   The other three were the Information Division (G2A), the Secret Service Division (G2B) and the Topography Division (G2C). (Gilbert, 2015, p.227)

Nolan and Charles H. Grasty, a well-known correspondent for the *New York Times*. The voyage was intended to be a heavily guarded military secret, but precautions were haphazardly applied. Thirty passengers forgot to arrive in civilian clothing and crates and boxes marked "War Department: Pershing Party, A.E.F., SS. Baltic, Chelsea Pier 23" sat on a dock for several days prior to boarding (Crozier, 1959, p. 124). Journalistic restraint and perhaps an element of luck prevented the voyage's details from reaching the American press or from being transmitted to Germany by enemy agents. *The Baltic* arrived in Liverpool ten days later. There, Floyd Gibbons, a correspondent, circumvented censorship rules by reporting back to the *Chicago Tribune* that Pershing had received a "Hearty welcome by the Mayor of Liverpool." These events demonstrate both the AEF's disorganization during its early existence and the rebelliousness of its accompanying correspondents.

The *Baltic* arrived in Boulogne on June 13 and Palmer visited the French Press Bureau located on Rue de Grenelle in Paris. He was unimpressed. He described the censors as "autocrats" who had the "power of life and death over words." Furthermore, he found the atmosphere in the office to be stale and the décor to be dingy (Casey, ND. Ch. 2, p. 3). The French War Ministry since August 1914 had required all publications throughout the country to submit copy to both local and Parisian censors prior to publication. The censors whom Palmer observed had worked day and night for almost three years in a confined office space.

Palmer understood he would have to work closely with the French. All telegrams sent from France were subject to French censorship as were all American newspapers publishing foreign editions in Paris. Pershing was soon able to negotiate an arrangement that allowed the AEF to transmit messages related specifically to the AEF under the condition that they did not discuss the French Army (Casey, ND. Ch. 2, p. 2). American newspapers published in France, however, would remain censored by the French.

Palmer set up a small Parisian office at Castellane House, 31 Rue de Constantine. His first few weeks were hectic. In addition to supervising the refurbishment of his new office, he hired censors, correspondents, a secretary, drivers and other G2D support staff. Pershing decided, with Washington's approval, that the AEF would officially accredit 12 correspondents who would primarily be selected from the AP, the UP and the INS.[2] The accredited news agencies and newspapers would pay a 3,000 dollar deposit to the War Department, which would be forfeited upon violation of AEF regulations. Palmer interviewed roughly 100 applicants for the correspondent positions, not all of whom were professional journalists. It was during this time that Palmer hired Joseph C. Green as Chief Field Officer. Until January 1918, Green played an important disciplinary role by enforcing censorship instructions and by occasionally imposing sanctions for improper conduct. (Crozier, 1959, pp. 126-128) His officious manner and media inexperience later caused tension between Palmer's office and the correspondents, many of whom were young and balked at military discipline.

Coordinating with the Allies

On June 25, in preparation for the arrival of the first AEF contingent to France, Palmer issued G2D's first list of censorship rules in a secret meeting with correspondents from the *New York Times,* the *New York Herald,* the *Associated Press* and the *United Press.* The attendees were prohibited from mentioning:

1. The name or location of the port of disembarkation, or the names of enemy units.
2. Names of any officers except commanders of divisions or the Commander-in-Chief, or the names of any units.

---

[2] In comparison to the French who fielded a much larger army also had 12. The British had only 5. (Crozier, 1959, p. 126-127)

3. Anything which will in any way indicate to the enemy the route of transports or the methods of the navy in safeguarding their passage.

Everyone in the room agreed not to mention the troops' arrival until after all 14,000 men had safely disembarked (Crozier, 1959, p.129). Several technical mistakes and miscommunications, however, allowed the location of St. Nazaire to be named as a point of disembarkment in the French press while the operation was still in progress. The slipup led to the first confrontation between G2D and its Allied counterparts and drove the AEF to pursue a more independent approach toward press censorship.

On June 19, the AEF instructed the French not to allow the publication of news related to the landing (Censorship registry, 19 June 1917, F rés 0270 C, BDIC). The French Press Bureau responded by issuing orders on the 22nd and 23rd to all French newspaper editors that "regarding the arrival of the Americans in our ports, names of boats, dates of arrival, names of arrival ports and histories of confrontations with submarines are forbidden. Only the lists of the Americans arriving are allowed to be published" (Censorship registry, 22 June 1917, F rés 0270 C, BDIC). Although the message was repeated for the next five days, it is not recorded as having reached the French telegraph agency (Telegraph censorship registry, June, 1917, F rés 0270 TAC, BDIC). On June 26, Palmer stressed the importance of the order to Captain Ribouillet, Director of the French Press Bureau, and reminded him again during the next two days (Censorship registry, 27-28 June 1917, F rés 0270 C, BDIC). June 26 was a busy day at the French Press Bureau and the censor responsible for passing the transmission had neglected to delete the line, "St. Nazaire June 26 (delayed)." A telegram was sent to Paris and slipped past the censors while another was successfully sent to London through the mail. The telegraph operator who passed the telegram probably had not been given the order to stop the information and may have been too overburdened to raise the matter to his superiors. News of the embarkment quickly spread to newspaper offices in Paris, London and New York (Crozier, 1959, p. 134).

On June 28, articles were printed in several New York and London papers mentioning that the first AEF corps had arrived in France. A particularly dangerous headline appeared in Paris - *Ce Soir* declared, "The first American Expeditionary Corps Has Arrived at St. Nazaire." The full story had been slashed by the French censors[3] but the headline, which revealed St. Nazaire as the port of debarkation, remained untouched. French police tried to retrieve all copies of the edition from newsstands but over 1000 were sold (Crozier, 1959, p. 132-133). Although French authorities suspended *Ce Soir* for three weeks (later shortened to 12 days), (Censorship Order, 28 June 1917, F rés 0270 CG BDIC, Censorship Registry, 28 Jun.1917, F rés 0270 C BDIC, 5N 499, 28 Jun.1917 Service Historique de la Défense) the incident damaged early Franco-American co-operation efforts over press censorship.

The danger posed to the American troops landing was slight. In June, 71 ships using five convoys departed from the American naval base at Hampton Roads and none were lost (though one was torpedoed in the Channel). The incident, however, led Palmer to establish a greater role for American personnel in censoring news related to the AEF. Here he achieved moderate success. Green was soon moved to the French Press Bureau, 110 Rue de Grenelle in Paris, where he had full authority to censor or pass any materials sent from American correspondents to American newspapers. He only had an advisory role, however, regarding news destined for French papers or telegrams sent to countries other than the United States (Casey, ND, Ch.2. p. 8 and McCabe, 1917). French censors monitored news written for French editions of American newspapers, even

---

3   Offending materials in French newspaper articles were crossed out by the censor's blue pencil prior to publication. As a result, French newspapers regularly ran stories containing large passages entirely blanked out.

in editions created specifically for American soldiers; a policy which was heavily criticized by American editors.

The mistake over St. Nazaire also led the AEF to push the French to more vigilantly protect Allied naval information. The AEF now insisted on a full prohibition on the naming of landing ports and all routes taken by naval and merchant ships. French naval censorship had been lax since 1914, much to London's chagrin.[4] Now the French Press Bureau made it a priority. On July 1, Pershing ordered that "everything of European origin pertaining to submarine warfare and to the convoying of vessels, so far that the American Navy was concerned" would be transmitted only with "the approval of the Naval Attaché at the American Embassy in Paris" (Pershing, 2017). The actions of the Allied navies and in particular the adoption of the convoy system in May 1917, prohibited the German U-Boat campaign from blocking the arrival of American troops to Europe in 1917 and 1918. Though it is impossible to fully gauge the extent to which Allied cooperation over naval censorship contributed to this feat, it surely played a positive role in its achievement (Sorrie, 2014, p. 175).

Over the summer, both the French and British were tightening their own military censorship controls and welcomed American enthusiasm for greater inter-allied co-operation. Pershing confronted the French in June over reports in the French media which exaggerated American fighting capabilities and suggested the immediacy in which the AEF's arrival to France would tilt the balance in the Allies' favor. If the Americans had reason to distrust the media in Paris and London, the French and British also had incentive to urge caution on American war correspondents. For the French, details of the failed Nivelle Offensive in late-May, followed by mutinies which, by the end of June, had affected up to 50 French divisions and as

---

4 The Official British Press Bureau Instructions outlines its orders regarding naval censorship and merchant shipping even before it discusses military matters. British Official Press Bureau Instructions, Box 6125 (5), Entry 221, AEF, GHQ, G2D, RG 120, NARA.

many as 30,000 soldiers, needed to be safeguarded from both the Germans and the increasingly strained French home front. The British were fighting what was becoming a ferocious campaign in Flanders and secrecy regarding British planning and maneuvers, it was feared, needed to be emphasized on the newly arrived and overly enthusiastic American reporters (Casey, ND, CH 2, pp. 11-12).

In July, Pershing ordered reporters not to mention the size of units, the AEF's future plans, new types of weapons, or the eventual position of the United States forces "on the firing line" (Casey, ND, CH 2, p. 13). In August, Secretary of War Newton D. Baker sent a list of similar orders to the British Press Bureau, which were relayed to both British censors and newspaper editors to protect military information related to the AEF. British censors and journalists now followed the same instructions as their American counterparts when handling AEF related news.

Palmer in Charge

On July 23, the AEF's accredited correspondents were moved to new lodgings at the Hôtel de la Providence in Neufchâteau—a small town approximately 150 miles east-southeast of Paris. The location was chosen for its proximity to the early AEF training grounds, its distance from the distractions of Parisian nightlife, and its two-hour travel time to GHQ which, Pershing believed, was far enough that reporters would not cause a nuisance by asking questions or requesting interviews.[5] For the next eight months, Palmer kept the reporters in Neufchâteau under close surveillance

---

5 Neufchâteau was a roughly one-hour drive to the AEF First Division training grounds in Gondrecourt, another hour's drive to Pershing's office in Chaumont and a half- day voyage to the AEF First Division's artillery training ground in Le Valdahon. See Crozier, 1959, p.142-143. Pershing's first interview with the reporters in August was a rare event. His disposition was both antagonistic and secretive

by monitoring their copy, travels and expense accounts. The boredom felt by most journalists in the provincial town was exacerbated by the slow pace of AEF mobilization, GHQ's reticence in its interactions with the press, and the frequent delays with which they received official war news compared to their British and French counterparts. The situation provided limited opportunity for imaginative journalism.

In September, American war correspondents were separated into two groups: accredited and visiting correspondents. Accredited correspondents were lodged in Neufchâteau and had regular access to AEF training grounds. Palmer believed that regular contact with soldiers would inspire comradery with the AEF and stress the importance of protecting military secrets (Casey, 2014, p. 21). To further embed these mostly young reporters with generally little or no military background, they were dressed in junior officers' uniforms. Visiting correspondences mostly lived in Paris and were occasionally granted guided tours of the training areas (Mander, 2010, p. 49). Both groups were required to submit all copies to censorship offices in Neufchâteau or Paris before being wired to the United States. All material destined to be shipped to the United States by post required a stamp issued by the censor's office.

In late summer to autumn 1917, the AEF expanded its presence at the French press bureau (relocated to the Parisian Bourse) and moved G2D's head office to nearby Rue St. Anne. Mark Watson, an artillery man with press experience, was appointed as Churchill's assistant. Watson, a cautious censor, later oversaw day-to-day affairs at G2D's Paris office in 1918. The censors' primary duties at the Bourse were to record "all news of importance in itself" and "to create brief but descriptive summaries" of all news articles or telegrams directly concerning the American Army. Telegrams which had been filed and passed already by the office in Neufchâteau or by the French or British Headquarters were to be registered. If telegrams sent from

Neufchâteau to the United States via Paris were considered suspicious, they were returned for reexamination. If messages from Paris or Neufchâteau required deletion or alteration, their authors were to be contacted immediately by phone or form letter (1918 Staff Manual, p.20, Censorship Instructions for Censors Folder, Box 6126 (5), Entry 221, AEF, GHQ, G2D, RG 120, NARA). Articles could not be altered by censors though full deletions were permitted (1918 Staff Manual). Censors immediately dealt with all received messages and could not pass material on to their replacements from a following shift. Visitors unaffiliated with the Censorship Room were prohibited.

Palmer argued that a censor's duty had both "positive" and "negative" functions. He believed G2D's role was to support and defend the AEF's mission in the United States as well as protect military secrets. He stressed the importance of a mutual understanding between journalists and censors:

> If a dispatch seems to have a bad tendency as written, he should appeal to the correspondent's patriotism and suggest that he rewrite it in a different spirit . . . Remember that five minutes spent in reasoning with a correspondent may mean a constructive result instead of a destructive. The correspondent wants copy; his paper wants copy. Through him we reach the public. Ours is an opportunity of influence if we use it. If we only delete copy then we are simply a post-office, not even a clearing house. Let the correspondent realize your own high purpose and the appeal will invariably succeed. Where it does not, then be firm. (Palmer, ND)

On October 21, selected units from the AEF's First Division entered the front lines for a little under two weeks (Casey, 2014, p. 18). The French prohibited all media from accompanying the soldiers—a decision that

Palmer did little to protest and which greatly angered his correspondents. The AEF's capture of its first German prisoner within a week of manning the trenches was potentially a big scoop for the news-starved correspondents, but the circumstances surrounding his death demanded caution: the prisoner, Leonard Hoffman, a 20-year old Bavarian mail orderly, had been bayoneted in the chest after raising his hands in surrender with a minor gunshot wound. Two days later, three American soldiers had their throats cut in a German night raid. Palmer chose to censor all details of the Hoffman story except for mentioning his death in an American field hospital and stopped the second story in its entirety. Palmer's prohibition of all news regarding the AEF's first deaths in the trenches demonstrated a willingness to repress his journalistic instincts while operating as Chief Press Officer (Smith, 1999).

By December, Pershing had received enough complaints about G2D from both reporters and from Washington (Crozier, 1959, p. 173) that he decided to replace Palmer. Since the St. Nazaire debacle, G2D had developed a reputation in some circles for ineptitude. Reporters complained that Palmer's supervision of their personal behavior and expenditures had been unnecessarily severe and his judgments on censorship too arbitrary. Palmer was first sent to Washington on a confidential mission. While there, he gave numerous speeches promoting G2D's activities and met with several prominent opponents of press censorship including Theodore Roosevelt. Upon his return to France, Palmer delivered directives from the War Department to Pershing and Nolan authorizing them to relax censorship regulations. This was his last act as Chief Press Officer.

G2D Reorganized

In January 1918, G2D was re-staffed. Palmer became Pershing's personal assistant and was replaced as Chief Press Officer by Gerald Morgan, a First

Lieutenant and an experienced correspondent. Churchill was replaced as head of G2D by Colonel W.C. Sweeney and Major E.R.W. McCabe replaced Chief Field Officer Green who was sent from Neufchâteau to work in the Paris office. Much of the tension between journalists and censors in Neufchâteau was soon abated. Both Morgan and McCabe had a better rapport with the correspondents than did their predecessors and the new censors at both Neufchâteau and Paris were seasoned journalists. In February an official AEF newspaper, *The Stars and Stripes,* was established with Guy T. Viskniskki as its first editor (Katz, 1921). It still exists today.

While G2D and the correspondents began to cooperate more smoothly, GHQ still only reluctantly released official information. Three months after the First Division's brief occupation of a quiet sector along the front, it was placed into another location in Lorraine. Nolan successfully lobbied the French to allow American reporters to accompany the First Division but once reporters arrived French officers refused them access. After 17 days, GHQ posted an official announcement, "American officers have been authorized to state that the sector of the Western Front taken over by the US Army is in the Lorraine, to the Northwest of Toul." Because of the information delay from GHQ and restrictions imposed by the French, American audience correspondents received the news more than two weeks after their British and French counterparts (Crozier, 1959, pp. 189-190). Although it had recently opened a second press headquarters in Beauvais, G2D still had not established a reliable method to extract information from GHQ or from America's allies.

In the months preceding the AEF's tremendous influx of troops to France, G2D suspended the accreditations of three correspondents who had publicly protested its censorship regime. Wythe Williams of the *New York Times* and *Collier's Weekly* was a prominent member of the American War Publicity League in France—a group of journalists who lobbied Washington to relax censorship regulations and had a substantial role in Palmer's replacement in January. Williams argued early on that the military should be removed entirely from the censorship process, and in February sent an article to *Collier's* blaming General Robert Nivelle (French Commander-in-Chief December 1916 – May 1917) for the French failure on the Chemin des Dâmes. His accreditation was suspended on February 26 (Crozier, 1959, pp. 179, 193, 280). Heywood Broun of the *New York Tribune* was suspended the next day. Since the fall, Broun had not only admitted to evading censorship on multiple occasions by mailing articles to the United States without submitting them to censorship but had confided to Green and Viskniskki that he felt unbounded by the terms of his contract with G2D. Broun's circumvention of censorship regulations and his harsh criticism of the AEF's generals had intensified since January and his suspension served as an example to other correspondents. Reginald Wright Kaufman of the *Philadelphia North American* was the third suspended reporter. Like Broun, he had found ways to evade censorship since the fall of 1917. Theodore Roosevelt had helped Kaufman circumvent G2D by receiving his messages in the mail and delivering them to the *Philadelphia North American*. By April, Military Intelligence followed Kaufman's activities in Paris, and both Secretary of State Robert Lansing and the War Department wanted his credentials suspended. G2D obliged on April 28, 1918. G2D had now removed whom it considered to be its least dependable correspondents. The timing was important.

In February and March 1918, the Doughboys fired roughly 1000 and 3000 shells respectively at the German lines and suffered over 500 losses under enemy fire. In early March, the British and French convinced the AEF to release casualty figures in staggered groups (as opposed to announcing them as they were collected) to the Committee on Public Information (CPI)[6]

---

6 America's domestic propaganda agency from April 1917 and its main propaganda organ in Europe during the war's final phase.

which was in turn responsible for distributing them to American newspapers. This staggered method had been employed already by the British and French for over a year and was intended to mask the results of enemy gas attacks and artillery barrages from the Germans (Casey, 2014, p. 22).

On March 21st, Germany launched its first of five attacks, collectively known as the Ludendorff Offensives, which lasted until July 18. On April 2, while the Germans inflicted high casualties and swiftly penetrated the Allied front, Pershing and Baker decided to ban officials within the United States from releasing information on the AEF. This panicked decision led to a total suspension of casualty announcements during one of the war's deadliest phases. Major newspaper editors bitterly complained. The *New York Times,* for example, fumed that Americans would not tolerate "an arbitrary and stupid censorship" (Casey, 2014, p. 24). Republicans in Washington insisted that the AEF was using censorship to mask its incompetence. Rumors soon spread about the sudden loss of 250,000 casualties by an unspecified Allied army at the very start of the offensive (the British had suffered 200,000 casualties during the offensive's first two weeks) though the AEF First Division had been held in reserve rather than face the advancing Germans directly. Pershing was forced to reverse the blanket ban within a week of its implementation (Stevenson, 2011, p. 55).

In April, Sweeney circulated a memo which outlined G2D censorship's basic guidelines, codified its existing rules and loosened several restrictions. The "Principle of censorship," it declared, was that "all information which is not helpful to the enemy may be given to the public." Four conditions were to be met by all articles: "1. They must be accurate in statement and implication. 2. They must not supply military information to the enemy. 3. They must not injure morale in our forces here, or at home, or among our allies. 4. They must not embarrass the United States or her Allies or any neutral countries." Names of places and names of AEF personnel or units were prohibited except under exceptional circumstances. All troop or ship movements were not to be discussed nor were the results of enemy fire. Finally, all exaggerations were to be avoided and casualty figures could only be passed as indicated in communiqués (Sweeney, 1918). These principles largely dictated G2D's approach to press censorship until the armistice.

In late April and early May, the War Department began to press Pershing to be more forthcoming in providing news (McCain Memo, 1918) and on May 10th, GHQ began to issue daily communiqués at 9PM. This was a useful step, but by the end of the month, the AEF had launched its first offensives and its increasing troop numbers and combat roles demanded more significant changes to G2D. On the morning of May 28th, the AEF captured the tiny village of Cantigny in the Somme and took 200 prisoners. Three days later, the Third Division helped defend Château-Thierry against a German counterattack. Château-Thierry's successful defense provided a useful propaganda tool for the Allies. Both American and French newspapers overstated the AEF's role by declaring that the Americans had "saved" Paris.

The positive AEF experience at Château-Thierry revealed serious flaws in G2D's news distribution methods. Neufchâteau and Beauvais were far from the fighting and correspondents who obeyed orders by waiting at press headquarters for communiqués were deprived of useful information until after the battle had concluded. The four American journalists who directly witnessed the fighting were infuriated when G2D prohibited the transmission of all information related to Château-Thierry until six days after selected journalists from Beauvais were permitted to visit the site and interview the troops.[7] Eventually, a French communiqué describing the battle determined what could be published (Dubbs, 2017, pp. 224-225). After Château-Thierry, the Beauvais office was closed and Neufchâteau

---

7   Two of these arrived at Château-Thierry by coincidence. The other two purposefully travelled to the front

became significant only as a G2D transit center for the rest of the war. After being re-located to Paris for only one week, the American press office moved to Meaux on June 7th.

Competition for scoops, particularly fierce between the growing number of visiting correspondents, encouraged rule breaking. In June, several important innovations were implemented to encourage cooperation. First, G2D began to brief correspondents prior to AEF attacks. On June 6, the Second Division attacked Belleau Wood and the battle's news reporting was more organized and focused than at Château-Thierry. The increasingly well-informed correspondents cooperated during Belleau Wood to write the AEF's first human interest story. A loophole in censorship regulations allowed "the marines," who were divided into brigades, to be mentioned as a single unit whereas other divisions or corps could not be identified similarly. The marines dominated the headlines and the battle remains significant to the corps' traditions today (Crozier, 1959, p. 221). Two days after the battle began, Pershing sent a cable to Washington advising the adoption of a pooling system inspired by the British and French but suited to America's more geographically scattered and diverse newspaper audience (Casey, 2014, p. 26). He suggested that 25 accredited correspondents be stationed in Meaux and one additional accredited correspondent be attached to each Division. Journalists assigned at the divisional level would be selected from current visiting correspondents and, where possible, would be assigned to divisions comprised of soldiers with similar geographic backgrounds.

Pershing's suggestions initiated several important changes to G2D over the summer. Many of the restrictions applied to visiting correspondents were removed and new correspondents, both accredited and visiting, were hired (Crozier, 1959, pp. 225, 279). AEF press officers were assigned at the corps level and a new pooling system was adopted (Casey, 2014, p. 27). Accredited correspondents were no longer preoccupied by competing with rival journalists because all had access to the same information. Because unfiltered war news was transmitted by AEF press officers at least twice daily and then supplemented by GHQ communiqués,[8] reporters focused on personal interest stories, regional angles and on the "larger aspects" of the war's operations (Memorandum from GHQ to McCabe, 2018). The new changes resulted in more thorough and engaging journalism. Even the CPI, which often clashed with G2D and was distrusted by the AEF, championed the new adjustments because they enhanced propaganda capabilities.

The Hundred Days and the Meuse Argonne

On July 18th, the Allies launched a counterattack at the Second Battle of the Marne. The event initiated the war's final phase, known as the "Hundred Days," a period marked by consecutive Allied victories and advances. Immediately, the French stopped exaggerating American accomplishments in the news and began to downplay the AEF's importance. Pershing insisted that American correspondents avoid exaggerations to compensate and emphasized self-reliance during the AEF's approaching campaigns.

The AEF fought under its own command for the first time from September 12-15 at the Battle of St. Mihiel. In preparation, correspondents in Meaux, monitored by German intelligence, were required for several weeks to strictly follow their daily routines or to remain at headquarters. Again, Pershing reminded correspondents to avoid exaggerations and ordered, "Don't win the war on this one." Specifically, they were to avoid mentioning Metz or its potential capture. On September 10th, the correspondents from Meaux were transported to Nancy. The AEF accommodated the American media

---

8   GHQ's comminiques followed in general the lines of the official communiques issued from the headquarters of the Allied Armies in France. (Casey, 2014, p. 208)

at St. Mihiel far more than in previous occasions. It installed an additional telegraph wire from Nancy to Paris and even housed correspondents in a hotel from which they could observe some of the opening barrage (Crozier, 1959, p. 238).

Competition between reporters and logistical difficulties, however, remained obstacles to G2D's efforts. Fred Ferguson, a veteran correspondent from the UP, wrote a piece on St. Mihiel the night before the opening barrage that was based on information provided during Nolan's prebattle press briefing. Morgan approved the story and it was quickly sent off to American papers. During the battle itself, news distribution was slowed down because of a sudden glut of transmissions and little information was sent back to United States. The AEF was successful at St. Mihiel. Against a retreating German 5th Army, it captured 200 square miles of land and 13,000 prisoners (Stevenson, 2011, p. 129). American newspaper reporting during the battle itself was primarily based on Ferguson's hypothetical report. Luckily, the AEF's plans were successful, and the stories retrospectively appeared accurate.

The Meuse-Argonne Offensive was launched on September 26 and became the AEF's largest and bloodiest engagement. When it concluded in November, the Americans had suffered 122,000 casualties which included over 26,000 killed. These staggering figures, coupled by the intensity of the fighting and the size of the battlefield, tested G2D's censorship and newsgathering capabilities more than other events. It proved to be G2D's high point, however, thanks to the dutifulness of its 25 accredited correspondents, the professionalism of its increasingly seasoned censors, and to procedural improvements adopted over the summer.

Because the Meuse-Argonne's 20-mile wide front (Crozier, 1959, p. 238) limited the scope for poignant personal observations, the correspondents (lodged at a newly established press headquarters in Bar-leDuc) benefitted greatly from the new pooling system. Background briefings and corps-level press releases enabled reporters to contextualize the AEF's slow advance and high casualty figures to American newspaper audiences by emphasizing the difficult battle terrain and formidable German defenses. The pooling system also freed up time and allowed correspondents to write articles which placed the offensive within the broad Allied advance of the war's final phase (Casey, 2014, pp. 32-33).

The "Lost Battalion" story was a product of the pooling system. On September 29, Lieutenant Kidder Mead, a press officer attached to I Corps, anticipated that the 77th Division would become isolated within enemy lines and briefed the accredited correspondents at Bar-le-Duc. Four days later when the division was surrounded, the correspondents were ready to launch what became the Meuse-Argonne's most important human-interest story. When news arrived that the 77th Division had been called on to surrender, the correspondents devised an appropriate headline response for the newspapers: "Go to Hell!" American readers closely followed the story until the division was relieved on October 7 (Crozier, 1959, pp. 254-255). The story of the "Lost Battalion" remains important to American military folklore.

Unprecedentedly high casualty figures at the Meuse-Argonne threatened to damage morale at home. The AEF earlier had been accused of hiding casualty figures and of masking the war's brutality from the media. To avoid recrimination, casualty statistics were now printed daily in American newspapers and in late-October, 1,453 were listed in a single day (Casey, 2014, p. 31). Yet in spite of these high figures, the Meuse-Argonne Offensive's immediate impact upon morale within the United States was relatively negligible for at least three reasons. First, the American home front had increasingly hardened throughout 1918 and, by the time of the Meuse-Argonne Offensive, exhibited a form of government-sponsored war hysteria. Second, a week after the battle began, Berlin publicly requested

Wilson for a ceasefire. The resulting exchange of notes between the two governments in mid-October took place in public and were heavily reported in American newspapers. Third, the accredited correspondents were highly self-disciplined during the war's final phase. Their mutual agreement to self-censor and to propagate the AEF's mission reflected an increased embeddedness into the ranks since the previous winter (Casey, 2014, p. 32). By the war's end, the relationship between the AEF and America's war correspondents largely depended on "voluntary censorship" (Memo Nolan to Churchill, 1918).

On November 7, censorship failed to prevent a premature armistice announcement from reaching the American press. The mistake sparked celebrations throughout the country, some larger than those four days later. The events which led to the infamous miscommunication occurred in rapid succession. On November 6, the Wolf Agency (Germany's semi-official news bureau) announced that German delegates had left to discuss armistice terms with an Allied mission. The next day, in the early afternoon, a phone call was placed to the American embassy in Paris from a man representing himself as a high-level French official who claimed that an armistice had been signed at 11 o'clock that morning. The news quickly spread to Brest, America's principal naval base in France and a wirehead where messages could be transmitted to New York without being redirected through Paris. Without confirming the information with Paris, Admiral Henry Wilson, Commander of the US Naval Forces in France, relayed the information to *La Dépêche*—the main local newspaper. Wilson then granted a request from the UP's Roy Howard to wire the news back to New York, which he did with the help of Ensign Sellards, Wilson's personal aid.

For different reasons, Howard and Sellards deliberately circumvented Allied censorship when sending the wire. After double-checking for an update at the American Embassy, Howard wanted to claim the story for the

UP before it jammed the lines. Sellards, uninterested in censorship, wanted simply to execute his boss's orders (Crozier, 1959, p. 263). William Phillip Simms was the only UP man authorized to send cable messages via Brest and Howard instructed a local operator to add Simms' name to the memo without his prior knowledge. Later, Sellards, after telling Howard to wait in the hallway, gave the message to another operator while the local censors were absent.

It not only contained the biggest news story since the start of the war but was signed by the UP's two most well-known figures in France. The operator assumed that such an important message had already passed through the Parisian censorship office and stamped it accordingly before sending it to New York. The news spread rapidly throughout the United States but nowhere else and the mistake proved to be more embarrassing than harmful. Wilson took the blame, publicly apologized and protected his aide. Howard enjoyed a successful post-war journalistic career working alongside Simms.

A Positive Force

The AEF's press censorship in Europe then was not entirely failsafe by the time of the armistice. Determined journalists could still, given the right circumstances, break the rules to get a jump on their competitors. During the war's final month, however, this happened far less often for three reasons. First, correspondents, particularly after the beginning of the Meuse-Argonne Offensive, travelled less outside the fighting zone and were therefore continuously monitored and scrutinized. Second, the correspondents most capable of deceit had been dismissed during the first half of 1918; their replacements were more reliable and those that remained fell into line. Finally, military-press relations had steadily improved throughout 1918. The

AEF increasingly trusted its correspondents with sensitive information and journalists in return voluntarily self-censored and propagated its missions and activities.

In conjunction with America's allies, the AEF developed a press censorship system in under 18 months whose effectiveness matched that of any other belligerent. As a subsection of the AEF, G2D's bureaucracy expanded greatly in mid-1918 as the number of both American troops and correspondents in Europe multiplied. In 1917, Frederick Palmer did a commendable job of building G2D bureaucratic foundations from scratch. However, it was in 1918 that significant adjustments to the AEF's news-gathering methods were introduced to compensate for America's increased combat role in France. By the fall, the High Command, censors and correspondents worked together under a system that created far less friction between them than before. When it was needed the most, G2D reached its pinnacle of effectiveness.

Immediately after the war, the changing political situation in Washington was unfavorable to the legacies of both G2D and the CPI. But by the time World War II began, both political and military circles closely studied the lessons learned from the World War I censorship experience. The Meuse-Argonne was and still is the deadliest battle fought by the American military, but it was also a moment where America's military censorship system operated smoothly to protect secrets, calm the home front, and encourage the troops. The fact that G2D's system was successfully built upon during World War II and Korea is perhaps its greatest historical vindication and legacy.

References

British Official Press Bureau Instructions, Box 6125 (5), Entry 221, AEF, GHQ, G2D, RG 120, NARA.

Casey, S. (N.D.) *"Censoring the Great War: World War I and the Birth of America's Modern Military-Media Relations."* Unpublished manuscript.

Casey, S. (2014). *When Soldiers Fall: How Americans Have Confronted Combat Losses from World War I to Afghanistan.* Oxford: Oxford University Press.

Censorship registry, 19 June 1917, F rés 0270 C, Bibliothèque de Documentation Internationale Contemporaine (BDIC).

Censorship Order, 28 June 1917, F rés 0270 CG BDIC, Censorship Registry, 28 Jun.1917, F rés 0270 C BDIC, 5N 499, 28 Jun.1917 Service Historique de la Défense (SHD).

Crozier, E. (1959). *American Reporters on the Western Front, 1914-1918.* Oxford: Oxford University Press.

Dubbs, C. (2017). *American Journalists in the Great War: Rewriting the Rules of Reporting.* Lincoln: University of Nebraska Press.

Gilbert, J. (2015). World War I and the Origins of U.S. Military Intelligence. Lanham: Rowman & Littlefield.

Katz, H.L. (1921) *A Brief History of the Stars and Stripes, Official Newspaper of the American Expeditionary Forces in France.* Washington DC: Columbia Publishing Company.

Mander, M. (2010). *Pen and Sword: American War Correspondents, 1898-1975.* Urbana: University of Chicago Press.

Memorandum by Major Palmer, Undated, Instructions for Censorship Officers Folder, Box 6127 (6), Entry 221, AEF, GHQ, G2D, RG 120, NARA.

Memorandum by General Pershing to American Military Attaché, London, England, 31 October 2017, Censorship of Naval Dispatches Folder, Box 6130 (9), Entry 221, AEF, GHQ, G2D, RG 120, NARA.

Memorandum by Colonel E.R.W. McCabe, Instructions for Censors' Officer folder, August 1917. Box 6127 (6), Entry 221, AEF, GHQ, G2D, RG 120, National Archives and Records Administration (NARA).

Memorandum to Colonel Nolan from Capt. Mark S. Watson, Assistant to the Chief, G2D, 10 April 1918, Box 6110 (1), Entry 222, AEF, GHQ, G2D, RG 120, NARA.

Memorandum from Lt. Colonel W.C. Sweeney, April 1918. Organization and Rules Folder, Box 6129, (8), Entry 221, AEF, GHQ, G2D, RG 120, NARA.

Memorandum by Major General H.G. McCain to Pershing, 23 May 1918. Complaints on Censorship Folder, Box 6130, (9), Entry 1241, AEF, G2D, RG 120, NARA.

Memorandum from GHQ to McCabe, July 1918, Press Section Liaison with Corps Folder, Box 6130 (9), Entry 1241, AEF, GHQ, G2D, RG 120 NARA.

Memorandum by Colonel McCabe on "Bourse Censors", August 1918, Box 6127 (6), Entry 221, AEF, GHQ, G2D, RG 120, NARA.

Memorandum by Nolan to Churchill, 30 October 1918, Complaints on Censorship folder, Box 6130 (9), Entry 1241, Entry 221, AEF, GHQ, G2D, RG 120, NARA.

Smith J.A. (1999). War & Press Freedom: The Problem of Prerogative Power. New York: Oxford University Press.

Sorrie, C. (2014) *"Censorship of the Press in France 1917-1918"*, Doctoral Thesis, London School of Economics.

Staff Manual, 1918, Censorship Instructions for Censors Folder, Box 6126 (5), Entry 221, AEF, GHQ, G2D, RG 120, NARA.

Stevenson, D. (2011). *With Our Backs to the Wall: Victory and Defeat in 1918*. London: Penguin.

Telegraph censorship registry, June, 1917, F rés 0270 TAC, BDIC.

Contributor Biographies

Jonathan A. Beall is an associate professor of history at the University of North Georgia. He earned his Ph.D. at Texas A&M in 2014 and specializes in military history. He teaches in UNG's Strategic and Security Studies Program.

Ashlee Beazley is currently a PhD Student and Teaching Assistant at the KU Leuven (Catholic University of Leuven), Belgium, where she is working on a thesis in comparative criminal law assessing the quality of legal assistance given to the defense in criminal proceedings. She holds a M.St. in British and European legal history from the University of Oxford, and a B.A. (History) and LL.B. (Hons.) degrees from the University of Auckland, New Zealand. Her research interests include legal history, 20th century history, criminal law and procedure, and criminal justice.

Terri Blom Crocker has a BA from Elmira College and a Master's and PhD from the University of Kentucky, where she was an instructor. She was also the Director of Investigations in the university's Office of Legal Counsel. She is now retired and lives in Fort Collins, Colorado.

Keith D. Dickson is Professor Emeritus, National Defense University. For over 12 years, he was Professor of Military Studies at the Joint Advanced Warfighting School at the Joint Forces Staff College.

Thomas I. Faith is a historian at the U.S. Department of State, and is the author of *Behind the Gas Mask: The U.S. Chemical Warfare Service in War and Peace*. He is a former Science History Institute fellow, and was the history and political science teacher at the U.S. House of Representatives Page School from 2005 to 2011. His fall 2016 article in *The Historian*, "It Would be Very Well if we Could Avoid it: General Pershing and Chemical Warfare" earned the Charles Thompson Prize from the Society for History in the Federal Government, and the Distinguished Writing Award from the Army Historical Foundation. He earned his PhD from George Washington University in 2008.

Shawn McAvoy is Associate Professor of History and Religion at Patrick & Henry Community College in Martinsville, Virginia, and has served as a Faculty Associate at Arizona State University in Tempe, Arizona. He publishes in both American History, where he focuses on the period from Reconstruction to the end of World War I, and in Religious Studies. His most recent articles are ""To Enlarge the Peaceful Influence of American Ideals": The Grant Administration's Optimism vis-à-vis German Unification in 1871" in the *Revista Romana de Studii Eurasiatice*, and ""We should not expect great benefit from America.": Japanese Expansion and the Breakdown of Communication within the Wilson Administration in 1914" in the *Journal of Asia-Pacific Studies*. McAvoy's current research on relations between Freedmen, the Freedmen's Bureau, and white citizens in Hanover County, Virginia, during Reconstruction is being worked into a monograph.

Jonathan S. Miner is a Professor of Political Science at the University of North Georgia. Dr. Miner is a scholar of International Relations, specializing in United States Foreign Policy, International Law, and Middle Eastern Politics, earning his J.D. from Drake University, M.A. in Political Science from the University of Iowa, and Ph.D in International Studies from the University of South Carolina. His dissertation, "Spokes of a wheel? Assessing combined efforts of government and civil society to stop international terrorism in the United States, Indonesia, Turkey, Spain and Russia" was completed in 2007. Courses Dr. Miner teaches include International Law, Foreign Policy Process, Global Issues, Middle East Politics, and Research Methods; he has led study abroad to Istanbul, Turkey (2010-2015) and directs the Model United Nations program at UNG.

B.D. Mowell is on the faculty of the School of Security and Global Studies at American Military University. He teaches undergraduate, masters and doctoral courses related to international relations, international organizations, and international security. To date, he has over 100 publications, principally related to international studies and security. His Ph.D. is in political science (international/comparative politics) from Florida International University and he completed post-doctoral work via Harvard University's Extension/Summer School from which he received a graduate certificate in international security.

Seyed Hamidreza Serri is an Associate Professor of Security Studies at the University of North Georgia. His scholarly works focus on Strategic &

Security Studies, Foreign Policy Analysis, and Quantitative & Qualitative Text Analysis. His recent works include "Operational Code Analysis: a Method for Measuring the Strategic Culture" in Stephen Walker and Mark Schafer (Eds.), *Operational Code Analysis and Foreign Policy Roles: Crossing Simon's Bridge*; and "Security and Military Power" in Dlynn Williams and Raluca Viman-Miller (Eds.), *Basics of World Politics*.

Charles Sorrie teaches at the Lycée Français de San Francisco. He holds a Doctorate from the London School of Economics and collaborated on the "Living Legacy of the First World War" project at the Carnegie Council for Ethics in International Affairs.

Raluca Viman-Miller is an associate professor at University of North Georgia, Georgia USA. She has a PhD from Georgia State University, a master's degree from Georgia Southern University and she completed her undergraduate work at "Babes-Bolyai" University in Cluj-Napoca, Romania. She has completed research and published on issues such as: the impact of migration on political behavior, political communication, regional and bilateral relations with implications on the European security. She teaches classes on Global Issues, Comparative Politics, European Politics, International Relations and American Government at undergraduate and graduate level.

www.ingramcontent.com/pod-product-compliance
Lightning Source LLC
Chambersburg PA
CBHW050103170426
43198CB00014B/2443